BURLEIGH DODDS SCIENCE: INSTANT INSIGHTS

NUMBER 09

African swine fever

Published by Burleigh Dodds Science Publishing Limited
82 High Street, Sawston, Cambridge CB22 3HJ, UK
www.bdspublishing.com

Burleigh Dodds Science Publishing, 1518 Walnut Street, Suite 900, Philadelphia, PA 19102-3406, USA

First published 2023 by Burleigh Dodds Science Publishing Limited
© Burleigh Dodds Science Publishing, 2024, except the following: Chapter 4 was prepared by a
U.S. Department of Agriculture employee as part of their official duties and is therefore in the public
domain. All rights reserved.

British Library Cataloguing in Publication Data
A catalogue record for this book is available from the British Library

ISBN 978-1-78676-861-2 (Print)
ISBN 978-1-78676-862-9 (ePub)

DOI: 10.19103/9781786768629

Typeset by Deanta Global Publishing Services, Dublin, Ireland

Contents

Series list

Introduction

African swine fever (ASF) is a viral disease affecting both wild and domestic pigs that was first identified in East Africa at the beginning of the 20th Century with the first scientific paper on the disease published in 1921 (Montgomery, 1921). Since this first reported outbreak, ASF has continued to spread across the African continent, with singular or recurring outbreaks recorded in 32 of the continent's 54 countries (World Organisation for Animal Health, 2022). However, it wasn't until 1957 that the disease reached Europe. The ASF genotype I (ESAC-WA genotype) was identified in a herd of pigs near Lisbon airport after they had consumed waste originating from international airline flights from Africa (Boinas et al., 2011). The 1957 outbreak of ASF was quickly eradicated. However it re-emerged in Lisbon in 1960 and spread throughout several other European countries, including Spain, Italy, France, Malta, Belgium, The Netherlands and Sardinia (EFSA Panel on Animal Health and Welfare, 2010). As a result of the implementation of strict biosecurity protocols and improved education surrounding disease transmission and prevention, ASF was successfully eradicated from Europe in 1995, with the exception of Sardinia where the disease remains endemic (Cwynar et al., 2019; Zhang et al., 2023).

Additional outbreaks of ASF were reported in South America and the Caribbean region in the 1970s, with the following countries recording outbreaks: Cuba, Haiti, Dominican Republic and Brazil (Costard et al., 2009). The last reported occurrence of the disease in these regions was in Haiti in 1984 (Costard et al., 2009).

With the exception of an isolated outbreak of ASF in Portugal in 1999, ASF was declared as eradicated from all countries outside of Africa – excluding Sardinia where, as noted, the disease is endemic (Gaudreault et al., 2020; Zhang et al., 2023).

However, in 2007 ASF genotype II was identified in the Republic of Georgia, with over 19% of the country's domestic pig population infected and euthanised as a result (Cwynar et al., 2019). Following this first identification, numerous outbreaks of the disease were then reported across Europe in wild boar populations and domestic pigs in Russia, Ukraine, Belarus, Lithuania, Latvia, Estonia, Hungary, Czech Republic, Poland, Romania, Bulgaria, Moldova and Belgium (Cwynar et al., 2019).

Since 2007 the disease has continued to spread across Europe and into Asia, with the People's Republic of China recording its first ever outbreak of ASF in 2018 (Gaudreault et al., 2020). Neighbouring countries, including Mongolia, Vietnam and Cambodia, subsequently reported devastating outbreaks within a year. This most recent outbreak of ASF is estimated to have resulted in the death of over 43 million pigs either as a result of contracting the disease or being culled as a precautionary measure to prevent further spread (You et al., 2021).

With no effective vaccine currently available, coupled with the virus' high mortality rate (up to 100%) (Schulz, 2019), controlling the spread and transmission of the disease has proven to be an extremely difficult task. The World Organisation for Animal Health (WOAH) 2022 report titled *African Swine Fever (ASF) - Situation Report 11* suggests that, since 2005, ASF has been reported in 73 countries (World Organisation for Animal Health, 2022).

This guide synthesises and reviews the wealth of research on key aspects of understanding, tracking and preventing this devastating disease.

The first chapter looks at recent advances in understanding the characteristics and epidemiology of ASF. The chapter provides a brief overview of the distribution of the different strains of ASF before moving on to discuss the clinical signs of the disease, focussing on the hyperacute, acute, subacute and chronic symptoms of the disease. The chapter concludes with a discussion on the modes of disease transmission and host cycles.

The second chapter reviews recent advances in surveillance and diagnostic techniques for tracking the spread of ASF. The chapter begins with an overview of the direct and indirect detection methods currently available to laboratories, including polymerase chain reaction (PCR) and enzyme-linked immunosorbent assays (ELISA). A section on the use of alternative sample matrices for diagnosing ASF is provided. The chapter then considers the recent advances and possibilities for genomic epidemiology.

The third chapter provides an overview of the surveillance tools and methods currently available to aid the early detection of ASF in pigs, before moving on to review the range of biosecurity measures which can be implemented to prevent its further spread. The chapter highlights the importance of adequate cleaning and disinfection procedures, as well as the influence of farm biosecurity and good farming practice on preventing outbreaks of ASF.

The final chapter considers recent advances in developing an effective vaccine for ASF. The chapter includes a comprehensive overview of the recent research completed on vaccine development, citing published reports on subunit vaccines, naturally or cell culture passed viruses used as attenuated vaccine backbones, as well as live attenuated vaccines using targeted deletions in virulent ASF virus isolates.

References

Boinas, F. S., Wilson, A. J., Hutchings, G. H., Martins, C. and Dixon, L. J. 2011. The persistence of African swine fever virus in field-infected *Ornithodoros erraticus* during the ASF endemic period in Portugal. *PLoS ONE* 6(5), 1-5. https://doi.org/10.1371/journal.pone.0020383.

Costard, S., Wieland, B., de Glanville, W., Jori, F., Rowlands, R., Vosloo, W., Roger, F., Pfeiffer, D. U. and Dixon, L. K. 2009. African swine fever: How can global spread be prevented? *Philosophical Transactions of the Royal Society of London: Series B, Biological Sciences* 364(1530), 2683–2696. https://doi.org/10.1098/rstb.2009.0098.

Cwynar, P., Stojkov, J. and Wlazlak, K. 2019. African swine fever status in Europe. *Viruses* 11(4), 310. https://doi.org/10.3390/v11040310.

EFSA Panel on Animal Health and Welfare. 2010. Scientific opinion on African swine fever. *EFSA Journal* 8(3), 149. https://doi.org/10.2903/j.efsa.2010.1556.

Gaudreault, N. N., Madden, D. W., Wilson, W. C., Trujillo, J. D. and Richt, J. A. 2020. African swine fever virus: An emerging DNA arbovirus. *Frontiers in Veterinary Science* 7(215), 1–17. https://doi.org/10.3389/fvets.2020.00215.

Montgomery, R. E. 1921. On A form of swine fever occurring in British East Africa (Kenya Colony). *Journal of Comparative Pathology and Therapeutics* 34, 159–191. https://doi.org/10.1016/S0368-1742(21)80031-4.

Schulz, K., Conraths, F. J., Blome, S., Staubach, C. and Sauter-Louis, C. 2019. African swine fever: Fast and furious or slow and steady? *Viruses* 11(9), 866. https://doi.org/10.3390/v11090866.

World Organisation for Animal Health 2022. African swine fever (ASF). Situation Report 11. Available at: https://www.woah.org/app/uploads/2022/06/asf-report11.pdf (Accessed 25 May 2023).

You, S., Liu, T., Zhang, M., Zhao, X., Dong, Y., Wu, B., Wang, Y., Li, J., Wei, X. and Shi, B. 2021. African swine fever outbreaks in China led to gross domestic product and economic losses. *Nature Food* 2(10), 802–808. https://doi.org/10.1038/s43016-021-00362-1.

Zhang, H., Zhao, S., Zhang, H., Qin, Z., Shan, H. and Cai, X. 2023. Vaccines for African swine fever: An update. *Frontiers in Microbiology* 14, 1–19. https://doi.org/10.3389/fmicb.2023.1139494.

Acknowledgements

We wish to acknowledge the following for their help in reviewing particular chapters:

- Chapter 1: Prof. Karl Ståhl, National Veterinary Institute, Sweden
- Chapter 2: Dr Vittorio Guberti, ISPRA, Italy
- Chapter 3: Dr Erika Chenais, National Veterinary Institute, Sweden
- Chapter 4: Dr Chris Netherton, Pirbright Institute, UK and Prof. Jose Sánchez-Vizcaíno, Universidad Complutense de Madrid, Spain

Chapter 1

Advances in understanding the characteristics and epidemiology of African swine fever

Youming Wang and Lu Gao, China Animal Health and Epidemiology Centre (CAHEC), China

1 Introduction

The causative agent of African swine fever virus (ASFV) is a unique member of the Asfarviridae family. ASFV is a genetically enveloped virus containing complex double-stranded DNA up to 190kb in length. The virion has an icosahedral structure with a five-layer structure with an average diameter of 260–300 nm (Wang et al., 2019).

The nucleoid of the virus particle consists of the viral genome, parts of transcriptional enzymes and DNA-binding protein (p150, p37, p14, p14.5, p10, etc.) (Wang et al., 2021). Different strains have variable genome lengths (Dixon et al., 2013). The genome contains genes that can be used to distinguish genotypes (i.e. p72), as well as five multiple gene families (MGFs) that can be used for epidemiological analysis. MGFs of different isolates vary greatly and may explain the mechanism of virus antigen variation and the ability of the virus to evade the host defense system (Yoo et al., 2020).

The layer surrounding the nucleoid is a 180-nm-diameter core shell. Both pp220 and pp62 polyproteins are major components of the core shell and are expressed during late virus replication. The so-called inner membrane is about 70Å thick and consists of pE183R, p17, pE183L, p12, pE248R and pH108R.

http://dx.doi.org/10.19103/9781786768629.01

The capsid layer, which comprises p72, has an icosahedral symmetry and is wrapped in an outer membrane. The virus is still infectious in the absence of the outer membrane (Andrés et al., 2020). Virus particles contain enzymes such as deoxyuridine pyrophosphatase, DNA polymerase, AP endonucleases and protein kinases. The enzymes are required for nucleic acid metabolism, DNA replication and repair and mRNA transcription and processing (Dixon et al., 2004; Yoo et al., 2020).

2 Distribution of African swine fever

2.1 Spatial and temporal distribution of the disease

African swine fever (ASF) was first identified as distinct from classical swine fever in Kenya in 1921 when it was associated with warthogs (Montgomery, 1921). ASF is believed to have started with infection of common warthogs by argasid ticks of the *Ornithodorosmoubata* complex. This cycle is typically not characterized by overt disease or mortality. Outbreaks in domestic pigs have been traced back to the early twentieth century in Eastern Africa, e.g. Kenya (1900s), Tanzania and Zambia (1910s), Malawi (1930s) and Mozambique (1950s) (Mulumba–Mfumu et al., 2019). In Southern Africa, outbreaks in domestic pigs have been traced back to the 1920s in South Africa and Namibia. By the 1960s, ASF outbreaks had been reported in most (20) countries in Eastern and Southern Africa, with the disease also spreading to West and Central Africa. The epidemiology is complex with 24 p72 genotypes identified (Quembo et al., 2018).

ASF is likely to have been introduced into Portugal via infected pork from Angola (a former colony) in 1957. Although the initial outbreak was controlled, the disease was subsequently reintroduced in 1960 when it became endemic in Portugal and Spain. ASF then spread to other countries in Europe (e.g. France, Italy, Malta, Belgium, the Netherlands and the former Soviet Union) (Bosch et al., 2016; Sánchez-Vizcaíno et al., 2015). By 1995, the implementation of rigorous disease control programs led to the disease being eradicated in affected EU countries with the exception of the island of Sardinia (part of Italy) (Costard et al., 2013a).

More recently, a highly virulent ASFV strain belonging to genotype II was introduced to Georgia in 2007 (probably from Angola). It then spread throughout the Caucasus and the Russian Federation before moving into Eastern Europe, the Baltic States and then some Western and Southern European countries (Chenais et al., 2019; Gogin et al., 2013; OIE, 2022). A recent report by the European Food Safety Agency (EFSA) confirms cases of ASF during the period 2014–21 in the following EU countries: Belgium and Germany in Western Europe; Greece and Italy in Southern Europe; the Baltic States (Estonia, Latvia and Lithuania); Poland, Hungary, Slovakia, Bulgaria, Romania and Czechia in Eastern Europe. Cases were also reported by the EFSA in a number of non-EU states: Belarus, Moldova, Serbia, the Russian Federation, Ukraine and North Macedonia (Baños et al., 2022).

ASF spread from Spain to the Caribbean and South America in the 1970s, including Cuba (1971 and 1980), Dominican Republic (1978), Brazil (1978) and Haiti (1979); successful disease control programs eradicated the disease by 1981 (Sanchez-Vizcaino et al., 2012). Since the emergence of ASF in Georgia in 2007, the disease has also spread to many countries in the Eurasian region and then to East Asia. ASF spread rapidly within China in 2018 followed by countries in Southeast Asia such as South Korea, Laos, Vietnam, Cambodia, Indonesia, Myanmar and the Philippines. During 2020 and 2021, outbreaks were reported in countries such as India, Malaysia and Thailand (OIE, 2022).

2.2 Distribution of virus strains

As noted earlier, ASFV strains have been grouped into 24 genotypes on the basis of the B646L gene. Two have been sequenced: genotype I and genotype II (Njau et al., 2021). The long-term evolution of the virus among wildlife hosts means that all the genotypes can be detected in Eastern and Southern Africa. Genotype I is found widely in West and Central Africa. After the introduction of genotype I from Africa into Portugal and Cuba, it spread into Western Europe, the Caribbean and South America. It has been associated with lower levels of mortality and more frequent subclinical and chronic infection (Owolodun et al., 2010; Sanchez-Vizcaino et al., 2012). Genotype I has also been reported in Eastern Europe (Gallardo et al., 2019; Sánchez-Cordón et al., 2017; Zani et al., 2018).

Genotype II was transmitted from East Africa to Georgia in 2007 and, as noted, moved to the Caucasus, Eastern Europe and then into East and Southeast Asia. As the disease has evolved since its introduction in China in 2018, low virulent natural mutants were isolated in China during 2020. High-dose infections of these mutants are associated with increased mortality, while low-dose infection was associated with persistent, chronic disease (Sun et al., 2021). Surveillance in 2021 reported that genotype I strains are also prevalent in some areas of China. The phylogenetic analysis suggested the isolates belong to the same clade as the genotype I Portuguese isolates NH/P68 and OURT88/3. With genotype I infection, pigs have been shown to develop chronic signs of illness such as weight loss, intermittent fever, skin ulcers and joint swelling with more intermittent mortality (Sun et al., 2021).

Surveillance results indicate that three types of gene-deleted strains of genotype II are prevalent in China, including the double genes deletion strains of CD2v and MGFs, the CD2v-deleted strain and MGFs-deleted strain. These gene-deleted strains tend to cause milder, chronic disease symptoms.

Recently, mixed infections of genotype II and genotype I have also been detected in China. With the emergence of genotype I and mutant strains, the recombination of various strains will intensify, and such a complex epidemic situation will create new challenges for early diagnosis and disease control (Zhang et al., 2021).

3 Clinical signs and postmortem findings of African swine fever

Clinical signs of ASFV infection are highly variable depending on various factors such as virus virulence, swine breed, route of exposure, infectious dose and endemic status in a particular area (Daniel Beltrán-Alcrudo et al., 2017). ASFVs can be classified into three categories based on virulence: high, moderate and low virulence strains. Disease symptoms can vary from hyperacute to asymptomatic, with a wide range of mortality. Hyperacute disease induced by high-virulence isolates can cause 100% mortality. Acute forms can lead to 90-100% mortality, while subacute forms lead to 30-70% mortality. Attenuated viruses may result in chronic forms with mortality rates below 30%, with survivors recovering and becoming virus carriers.

3.1 Hyperacute symptoms

The hyperacute form is characterized by sudden death within 1-3 days after infection after a few clinical signs. High fever (41-42°C) may be seen in some infected animals before death. A few gross lesions may appear in organs together with the accumulation of fluid in the body.

3.2 Acute and subacute symptoms

In acute ASF, following an incubation period of 4-7 days (occasionally up to 14 days), high fever (41-42°C) will present combined with listlessness, reduced feed intake and labored breathing. Some neurological symptoms such as convulsion may occur (Fig. 1a). Skin flushing and, in particular, hemorrhage and cyanosis of the ears, legs and abdomen are more easily detected in white-skinned pigs than those with thick fur or dark skin (Fig. 1b). Vomiting, constipation, diarrhea, bloody feces and mucopurulent ocular and nasal secretions may be evident (Fig. 1b). Abortions may occur during pregnancy.

In the case of subacute infection, death may occur after some weeks with a high fatality rate in piglets. Fluctuant fever, swollen joints and secondary bacterial infection are usually present. Interstitial pneumonia may occur and lead to respiratory distress and coughing. Pregnant sows may abort.

Figure 1 Clinical signs of acute African swine fever. Source of Figure 1a: Chen et al. (2018).

An enlarged, friable, dark red to black spleen is the most recognizable postmortem finding (Fig. 2a). The lungs may display severe edema, with froth in the trachea and bronchus (Fig. 2b). Blood red peritoneal effusion may be present (Fig. 2c). The kidney and the renal papilla are swollen with faint yellow gel exudation (Fig. 2d). Enlarged, edematous and hemorrhagic areas can be observed in mesenteric, mandibular and inguinal lymph nodes, while the gastrohepatic and renally mph nodes show a marbled aspect (Fig. 3). Congestion and hemorrhages in serosa and hemorrhages in epicardium and endocardium may also be found. Petechiae, ecchymoses and hemorrhages in kidneys, intestines and urinary bladder may occur (Fig. 4) (Chen et al., 2018).

3.3 Chronic symptoms

Slight fever usually displays after 2-3 weeks of infection accompanied by mild respiratory distress and moderate-to-severe joint swelling. Skin erythema, necrosis and pneumonia have also been observed. Chronic disease is mainly characterized by respiratory tract lesions. Pneumonia with caseous necrosis (sometimes with focal mineralization) in the lungs, fibrinous pericarditis and lymphoreticuloendothelial hyperplasia often occur.

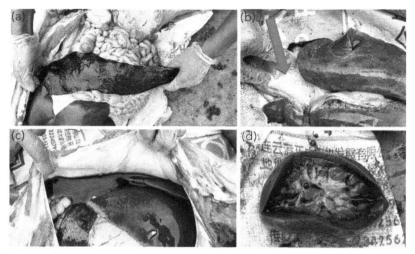

Figure 2 The most recognizable postmortem findings of acute African swine fever. Source: Chen et al. (2018).

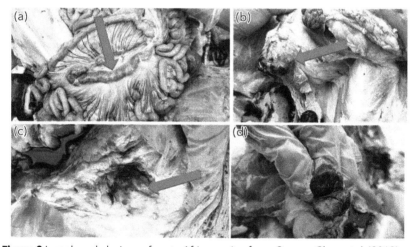

Figure 3 Lymph node lesions of acute African swine fever. Source: Chen et al. (2018).

4 Disease transmission: hosts and transmission cycles

4.1 Biological hosts

All members of the family Suidae are susceptible to African swine fever, especially domestic pigs. In Europe in particular, wild boar and feral pigs have been shown to have the same susceptibility to ASFV as domestic pigs (Jori and Bastos, 2009; McVicar et al., 1981). Warthogs (*Phacochoerus africanus*), bushpigs (*Potamochoerus larvatus*) and giant forest hog (*Hylochoerus*

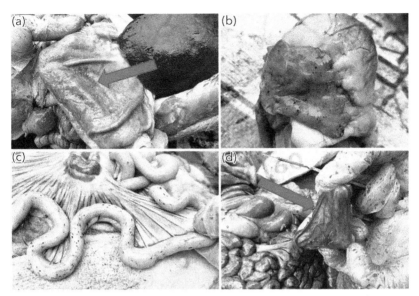

Figure 4 Hemorrhagic lesions of acute African swine fever. Source: Chen et al. (2018).

meinertzhageni) have shown resistance to ASFV and become asymptomatic carriers of the disease (Jori and Bastos, 2009).

Infected *Ornithodoros* ticks are the main biological vectors for ASFV, are able to retain the virus for several months or even years and then transmit it to susceptible hosts (Diaz et al., 2012; Rennie et al., 2010). It has also been suggested that the stable fly (*Stomoxyscalcitrans*) can transmit the virus within 48h after feeding on infected blood (Baldacchino et al., 2013; Mellor et al., 1987).

4.2 Transmission: sylvatic cycle

There are various transmission cycles which vary in importance in different areas. In Eastern and Southern Africa, ASF is maintained by an ancient cycle of infection between warthogs and *Ornithodoros* ticks known as the 'sylvatic cycle' (Costard et al., 2013b). Young suckling warthogs are infected in burrows infested with ASFV-positive soft ticks. These warthogs will develop a transient viremia which allows transmission of ASFV to naive ticks during blood meals (Plowright et al., 1969; Thomson, 1985). Ticks can amplify and maintain the virus for up to 15 months and transmit the virus to warthogs in the next farrowing season (Fernando et al., 2011; Hess et al., 1989; Ravaomanana et al., 2010). This ancient cycle only exists in part of southern and eastern Africa, and it is the origin of other cycles (Jori and Bastos, 2009).

4.3 Transmission: tick–pig cycle

Virus transmission from warthogs to free-range domestic pigs usually via *Ornithodoros* allows transmission into domestic pigs (Jori and Bastos, 2009). This acts as a bridge from the 'sylvatic cycle' to other transmission cycles. The so-called tick–pig cycle describes transmission from *Ornithodoros* ticks inhabiting pig sties to domestic pigs (Boinas et al., 2004; Sánchez-Vizcaíno et al., 2009).

4.4 Transmission: domestic cycle

Once the virus is introduced to domestic pig populations, transmission is possible via direct contacts and fomites. Human activities such as the movement of infected animals and the sale of infected pig products will facilitate virus spread (Morilla et al., 2002). The 'domestic cycle' is the primary transmission cycle of the disease in, e.g. China and Southeast Asia. It has been described in terms of inter-farm and intra-farm cycles (Fig. 5) (Gao et al., 2021).

Long-distance leaps of the virus are usually caused by the movement of infected pork products (Antonio et al., 2008; Botija and Badiola, 1966; Lyra, 2006; Wardley et al., 1983). Initial outbreaks in an area have been triggered by the illegal importation of infected or contaminated animal products (Costard et al., 2013a). Once these imported pork products have entered local markets, virus-infected kitchen waste may then cause infection among the domestic pig population. Once the disease is established in an area, virus spread may occur via three routes:

- First, slaughterhouses that process the infected pigs will be contaminated and produce virus-contaminated pork products. Once waste from these products is used as pig swill, the disease spreads.
- Second, infected pigs traded in live pig markets or between farms will result in the spread of the virus.
- Third, pig transport vehicles and personnel exposed to the virus will transmit the pathogen to ASF-free premises and trigger new outbreaks.

Intra-farm transmission occurs from physical contact between infected animals and susceptible individuals. Susceptible animals may also pick up the disease from infected blood or other excretions from sick animals (urine, feces) as well as contaminated equipment, vehicles, clothing and farm personnel (Gao et al., 2021; Sánchez-Vizcaíno et al., 2009; Mur et al., 2012).

In the domestic cycle, outbreaks related to feeding pigs with kitchen waste may dominate the initial stage of an ASF epidemic. These outbreaks mainly occur in small farms with poor biosecurity. Where swill feeding is better controlled, movement of sick pigs and mechanical transmission via personnel

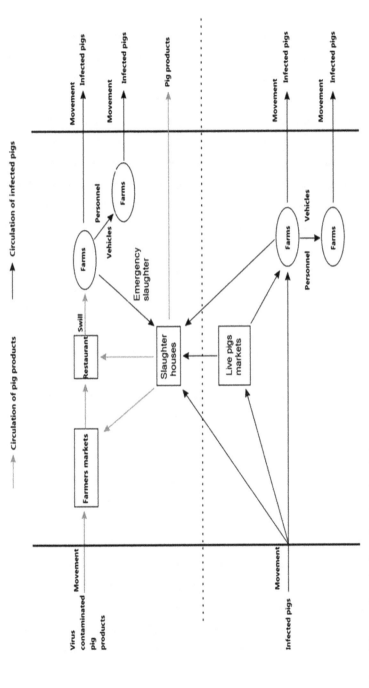

Figure 5 Domestic cycle along the pig value chain.

and vehicles become important mechanisms of transmission (Fig. 6). This can be seen in the analysis of the spread of the disease in China (Gao et al., 2021) (Tables 1 and 2). In poorer countries with a high breeding density, especially those with swill feeding, transmission among domestic pigs follows 'domestic cycle' characteristics (Antonio et al., 2008; Bora et al., 2020; Heilmann et al., 2020; Le et al., 2019; Lyra et al., 1986; Patil et al., 2020).

4.5 Transmission: wildboar habitat cycle

Virus from pig farms can be transmitted to wild boar populations in the 'wild-boar habitat cycle' (Chenais et al., 2018). This cycle is characterized by the continuous presence of the virus in the wild boar populations. Contamination is both from direct contact with infected animals and transmission from contact with ASFV-positive wild boar carcasses (Chenais et al., 2018). The long-term availability of the virus in infected carcasses enables the virus to persist irrespective of factors such as variation in the density of wild boar populations (Gogin et al., 2013; Pei et al., 2021). The wild boar habitat cycle has contributed to the persistence of ASF in Europe (Sauter-Louis et al., 2021).

5   Transmission dynamics of African swine fever

ASF transmission dynamics depend on factors such as hosts (domestic or wild animals), the features of the virus (strains and virulence), the environment and intervention measures (Schulz et al., 2019). Studies have suggested different latency periods. A number of studies suggest 4-6 days (de Carvalho Ferreira et al., 2013; Guinat et al., 2014, 2015; Pietschmann et al., 2015). Research on early outbreaks in China suggested an incubation period of 8.4-10.8 days (Li et al., 2021). The upper limit is 20 days. These estimated latent periods are shorter than the range defined by 'Terrestrial Animal Health Code' of OIE (World Organization for Animal Health) (15 days) (Oie, 2019) (Table 3).

One area of investigation has been rates of reproduction (as measured by the reproductive number R_0) with results assigned to different groups by the affected host (domestic/wild animals) and transmission route (within herds/between herds/free range) (Table 1). R_0 values for transmission between herds appear to be lower than those within herds but higher than those for free-range animals. ASF transmission within wild populations is generally slower than within domestic populations. Direct contact with infected animals or their excretions (especially infected blood) appears to pose the highest transmission risk (Schulz et al., 2019). The transmission risk shows a decreased trend with the declined frequency of effective contact. Variations in reproduction rates are probably the result of the limited number of experimental animals in particular studies (Table3). Comparing transmission dynamics with other highly contagious pig

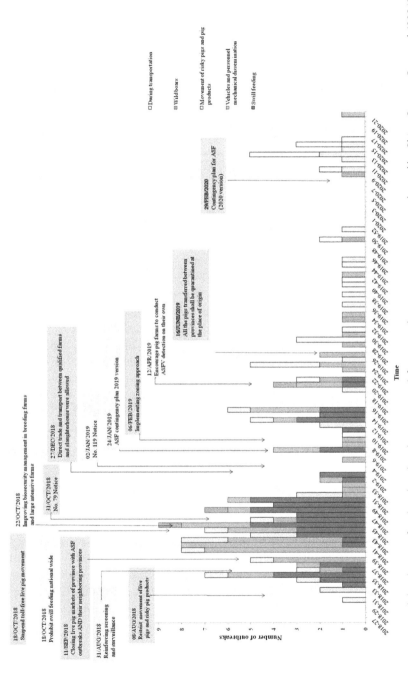

Figure 6 Temporal distribution of outbreaks related to various risk factors and intervention strategies adopted by China. Source: Gao et al. (2021).

Table 1 Risk factors of ASF transmission in various phases in China

Various phases	Risk factors		
	Swill feeding (%)	Movement of infected pigs and risk pig products (%)	Mechanical dissemination by contaminated vehicles, equipment and personnel (%)
First phase (−2018.9)	47	35	18
Second phase (−2019.8)	40	15	45
Third phase (−2020.4)	0	75	25

Source: Gao et al. (2021).

diseases, such as FMD and CSF, also confirms that ASF transmission is particularly dependent on intimate contact with infected animals or their excretions (e.g. blood, urine or feces) (Guberti et al., 2012; Hayer et al., 2017; Ster et al., 2012).

6 Conclusion

Since 2018, ASF has spread globally, with increased speed of transmission, especially in the Asia Pacific region. The epidemic characteristics have changed, including the emergence of mutant strains, mixed infections of different strains, and new transmission dynamics characteristics. ASF has had an enormously adverse impact on the swine industries across all production sectors, such as breeding, slaughtering, transportation, trading, and so on. In countries where the virus was newly introduced, the pathogen spilled over to the wild boar population as well, causing a certain impact on the wild boar ecology.

7 Where to look for further information

In recent years, the epidemic of African swine fever has shown new characteristics. It has been proposed that country-wide viral elimination of ASFV from newly infected countries is very unlikely to be achievable in the short to medium term. In view of this, the future research should focus on the following points. The first one is to study the etiological characteristics of new strains, which can be supportive for developing effective and safe ASF attenuated vaccine. On the other hand, developing the supportive techniques to achieve early detection, early diagnosis and early disposal, such as rapid tests and portable PCR will be good explorations. Last but not least, the research on transmission mechanisms of new strains in the newly infected countries will be helpful for disease control.

- The best source of information on global ASF can be acquired from the website of the FAO (https://www.fao.org/animal-health) and WOAH (https://

Table 2 The odds ratios of outbreaks occurred in different scale farms that related to the same risk factor in China

Risk factors	Farm size	Infected by this risk factor or not	Sample size	Odds ratio (95%CI)
Swill feeding	<500 heads	Yes	60	8 (1.7,36.3)
		No	67	
	501-2000 heads	Yes	3	2.1 (0.3,15.3)
		No	11	
	2001-5000 heads	Yes	0	NA
		No	9	
	>5000 heads (reference)	Yes	2	—
		No	16	
Risky pigs and pig production movement	<500 heads	Yes	25	1.2 (0.3,4.6)
		No	102	
	501-2000 heads	Yes	3	1.4 (0.2,8.1)
		No	11	
	2001-5000 heads	Yes	0	0 (NA)
		No	9	
	>5000 heads (reference)	Yes	3	—
		No	15	
Vehicles and personnel mechanical dissemination	<500 heads	Yes	42	0.2 (0.1,0.6)
		No	85	
	501-2000 heads	Yes	8	0.5 (0.1,2.2)
		No	6	
	2001-5000 heads	Yes	9	NA
		No	0	
	>5000 heads (reference)	Yes	13	—
		No	5	

Source: Gao et al. (2021).

www.woah.org/en/what-we-do/animal-health-and-welfare/animal -diseases/). The City University of Hong Kong (https://www.cityu.edu.hk/ jcc/) has carried out a lot of research with respect to the topic in this review, especially for the research on Asia Pacific region.

- There are a number of conferences designed to support ASF control and which identify current problems and ways of tackling them, including Standing Group of Experts on African swine fever under the GF-TADs umbrella, Meetings for FAO African Swine Fever Global Pool of Expertise.
- The Epidemiological analysis of ASF in the European Union countries can be acquired from EFSA Journal.

Table 3 Variation of latent period and calculated R_0 for ASF obtained from experimental and field studies

Study setting	Transmission probability	Type of the subjects	Strain	Latent period	Basic reproduction number (95% CI)	References
Experimental studies						
	Inoculation to all subjects	Domestic suids	Pol_28298_O111 isolate ('Georgia-like' strain) genotype II	5-20 days	–	Walczak et al. (2020)
	Inoculation to all subjects	Mixed herds of wild and domestic suids	'Armenia08' strain	4 days	6.1 (95%CI: 0.6-14.5)	Pietschmann et al. (2015)
	Inoculation to all subjects	Mixed herds of wild and domestic suids	'Armenia08' strain	4 days	5.0 (95% CI: 1.4-10.7)	Pietschmann et al. (2015)
	Between herds	Wild herds and domestic herds	'Armenia08' strain	4 days	0.5 (95% CI:0.1-1.3)	Pietschmann et al. (2015)
	Inoculation to half of subjects	Domestic suids	Malta'78	3-6 days	13.2 (95%CI:3.72-46.5)[a]	de Carvalho Ferreira et al. (2013)
	Inoculation to half of subjects	Domestic suids	Malta'78	3-6 days	24.2 (95%CI:8.99-64.9)[b]	de Carvalho Ferreira et al. (2013)
	Within herds	Domestic suids	Georgia 2007/1 ASFV strain	4.5 days	2.8 (95%CI:1.3-4.8)	Guinat et al. (2015)
	Between herds	Domestic suids	Georgia 2007/1 ASFV strain	4.5 days	1.4 (95%CI:0.6-2.4)	Guinat et al. (2015)
	Inoculation to all subjects	Domestic suids	Georgia 2007/1 ASFV strain	3.6-5.4 days	/	Guinat et al. (2014)

Field studies

Within herds	Domestic suids		10.00 days (95%CI:8.37,11.76)[c]	4.82 (95% CI: 3.84, 6.11)	Li et al. (2021)
Within herds	Domestic suids		8.38 days (95%CI:8.02,9.08)[c]	7.94 (95% CI: 7.26, 8.72)	Li et al. (2021)
Within herds	Domestic suids		10.76 days (95%CI:9.55,11.84)[c]	11.90 (95% CI: 10.71, 12.91)	Li et al. (2021)
Within herds	Domestic suids	Georgia 2007/1 strain		7.2 (95% CI:2.1,14.2)	Guinat et al. (2015)
Within herds	Domestic suids		5.8 days (95%CI:2.6–9.1)–9.7 (95%CI:6.5–12.9)	4.4 (95%CI:2.0–13.4)–17.3 (97%CI:3.5–45.5)	Guinat et al. (2018)
Between herds	Domestic suids	Georgia 2007/1 strain		3.5 (95% CI:1.2–7.0)	Guinat et al. (2015)
Free ranged	Domestic suids			1.58–3.24 (95%CI:3.21–3.27)	Barongo et al. (2015)
Wilderness survival	Wild suids			1.54(95%CI:1.11–2.37)	Jun-Sik et al. (2021)
Wilderness survival	Wild suids			1.139(95%CI:1.123–1.153)	Loi et al. (2020)
Wilderness survival	Wild suids			1.58(95%CI:1.13–3.77)	Iglesias et al. (2016)

[a]Group with high dose.
[b]Group with low dose.
[c]The value is the incubation periods, which include subclinical periods and latent periods.

8 References

Andrés, G., Charro, D., Matamoros, T., Dillard, R. S. and Abrescia, N. G. A. (2020), The cryo-EM structure of African swine fever virus unravels a unique architecture comprising two icosahedral protein capsids and two lipoprotein membranes, *J. Biol. Chem.*, 295(1), 1–12.

Antonio, A. E., Simeón-Negrín, R. and Frías-Lepoureau, M. (2008), Eradication of African swine fever in Cuba (1971 and 1980), *Trends in Emerging Viral Infections of Swine.*,125–131.

Baños, J. V., Boklund, A., Gogin, A., Gortázar, C., Guberti, V., Helyes, G., Kantere, M., Korytarova, D., Linden, A., Masiulis, M., Miteva, A., Neghirla, I., Oļševskis, E., Ostojic, S., Petr, S., Staubach, C., Thulke, H. H., Viltrop, A., Wozniakowski, G., Broglia, A., Abrahantes Cortiñas, J., Dhollander, S., Mur, L., Papanikolaou, A., Van der Stede, Y., Zancanaro, G. and Ståhl, K. (2022), Epidemiological analyses of African swine fever in the European Union, *EFSA J.*, 20(5),1–106.

Baldacchino, F., Muenworn, V., Desquesnes, M., Desoli, F., Charoenviriyaphap, T. and Duvallet, G. (2013), Transmission of pathogens by Stomoxys flies (Diptera, Muscidae): a review, *Parasite*, 20, 26.

Barongo, M. B., Ståhl, K., Bett, B., Bishop, R. P., Fèvre, E. M., Aliro, T., Okoth, E., Masembe, C., Knobel, D. and Ssematimba, A. (2015), Estimating the basic reproductive number (R0) for African swine fever virus (ASFV) transmission between pig herds in Uganda, *PLoS ONE*, 10(5), e0125842.

Boinas, F. S., Hutchings, G. H., Dixon, L. K. and Wilkinson, P. J. (2004), Characterization of pathogenic and non-pathogenic African swine fever virus isolates from Ornithodoros erraticus inhabiting pig premises in Portugal, *J. Gen. Virol.*, 85(8), 2177–2187.

Bora, M., Bora, D. P., Manu, M., Barman, N. N., Dutta, L. J., Kumar, P. P., Poovathikkal, S., Suresh, K. P. and Nimmanapalli, R. (2020), Assessment of risk factors of African Swine Fever in India: perspectives on future outbreaks and control strategies, *Pathogens*, 9(12).

Bosch, J., Rodríguez, A., Iglesias, I., Munoz, M. J., Jurado, C., Sánchez-Vizcaíno, J. and De, L. (2016), Update on the risk of introduction of African swine fever by wild boar into disease-free European Union countries, *Transbound. Emerg. Dis.*

Botija, C. S. and Badiola, C. (1966), Presencie of the African swine pest virus in Haematopinus suis, *Bull. Off.* Int. Épizoot., 66(1), 699.

Chen, C., Dong, Y., Kai, Y., Li, F., Jiang, X., Shan, Y., Fan, X. and Wu, X. (2018), The field diagnostic on first case of African swine fever in Jiangsu Province, *China Anim. Health Inspect.*

Chenais, E., Depner, K., Guberti, V., Dietze, K., Viltrop, A. and Ståhl, K. (2019), Epidemiological considerations on African swine fever in Europe 2014–2018, *Porc. Health Manag.*, 5(1), 6.

Chenais, E., Ståhl, K., Guberti, V. and Depner, K. (2018), Identification of wild boar-habitat epidemiologic cycle in African swine fever epizootic, *Emer. Infect. Dis.*, 24(4), 810–812.

Costard, S., Jones, B. A., Martínez-López, B., Mur, L., de la Torre, A., Martínez, M., Sánchez-Vizcaíno, F., Sánchez-Vizcaíno, J-M., Pfeiffer, D. U. and Wieland, B. (2013a), Introduction of African swine fever into the European Union through illegal importation of pork and pork products, *PLoS ONE*, 8(4), e61104.

Costard, S., Mur, L., Lubroth, J., Sanchez-Vizcaino, J. M. and Pfeiffer, D. U. (2013b), Epidemiology of African swine fever virus, *Virus Res.*, 173(1), 191-197.

Daniel Beltrán-Alcrudo, M. A., Gallardo, C., Kramer, S. A. and Penrith, M.-L. (2017), African swine fever detection and diagnosis-a manual for veterinarians, FAO Animal Production and Health Manual, Rome.

de Carvalho Ferreira, H. C., Backer, J. A., Weesendorp, E., Klinkenberg, D., Stegeman, J. A. and Loeffen, W. L. A. (2013), Transmission rate of African swine fever virus under experimental conditions, *Vet. Microbiol.*, 165(3-4), 296-304.

Diaz, A. V., Netherton, C. L., Dixon, L. K. and Wilson, A. J. (2012), African swine fever virus strain Georgia 2007/1 in Ornithodoros erraticus ticks, *Emer. Infect. Dis.*, 18(6), 1026-1028.

Dixon, L. K., Abrams, C. C., Bowick, G., Goatley, L. C., Kay-Jackson, P. C., Chapman, D., Liverani, E., Nix, R., Silk, R. and Zhang, F. (2004), African swine fever virus proteins involved in evading host defence systems, *Vet. Immunol. Immunopathol.*, 100(3-4), 117-134.

Dixon, L. K., Chapman, D. A. G., Netherton, C. L. and Upton, C. (2013), African Swine Fever Virus Replication and Genomics.

Fernando, B. N. T. W., Sudeshika, T. S. H., Hettiarachchi, T. W., Badurdeen, Z., Abeysekara, T. D. J., Abeysundara, H. T. K., Jayasinghe, S., Ranasighe, S. and Nanayakkara, N. (2011), Evaluation of biochemical profile of Chronic Kidney disease of uncertain etiology in Sri Lanka, *PLoS ONE*, 15(5)

Gallardo, C., Soler, A., Rodze, I., Nieto, R., Cano-Gómez, C., Fernandez-Pinero, J. and Arias, M. (2019), Attenuated and non-haemadsorbing (non-HAD) genotype II African swine fever virus (ASFV) isolated in Europe, Latvia 2017, *Transbound. Emerg. Dis.*, 66(3), 1399-1404.

Gao, L., Sun, X., Yang, H., Xu, Q., Li, J., Kang, J., Liu, P., Zhang, Y., Wang, Y. and Huang, B. (2021), Epidemic situation and control measures of African Swine Fever Outbreaks in China 2018-2020, *Transbound. Emerg. Dis.*, 68(5), 2676-2686.

Gogin, A., Gerasimov, V., Malogolovkin, A. and Kolbasov, D. (2013), African swine fever in the North Caucasus region and the Russian Federation in years 2007-2012, *Virus Res.*, 173(1), 198-203.

Guberti, V., Fenati, M., O'Flaherty, R. and Rutili, D. (2012), The epidemiology of Classical Swine Fever in the wild boar populations of Europe. STOP – CSF, International Scientific Conference(vi) Sad, Serbia.

Guinat, C., Gubbins, S., Vergne, T., Gonzales, J. L., Dixon, L. and Pfeiffer, D. U. (2015), Experimental pig-to-pig transmission dynamics for African swine fever virus, Georgia 2007/1 strain, *Epidemiol. Infect.*, 144(1), 25-34.

Guinat, C., Porphyre, T., Gogin, A., Dixon, L., Pfeiffer, D. U. and Gubbins, S. (2018), Inferring within-herd transmission parameters for African swine fever virus using mortality data from outbreaks in the Russian Federation, *Transbound. Emerg. Dis.*, 65(2), e264-e271.

Guinat, C., Reis, A. L., Netherton, C. L., Goatley, L., Pfeiffer, D. U. and Dixon, L. (2014), Dynamics of African swine fever virus shedding and excretion in domestic pigs infected by intramuscular inoculation and contact transmission, *Vet. Res.*, 45(1), 93.

Hayer, S. S., VanderWaal, K., Ranjan, R., Biswal, J. K., Subramaniam, S., Mohapatra, J. K., Sharma, G. K., Rout, M., Dash, B. B., Das, B., Prusty, B. R., Sharma, A. K., Stenfeldt, C., Perez, A., Delgado, A. H., Sharma, M. K., Rodriguez, L. L., Pattnaik, B. and Arzt, J.

(2017), Foot-and-mouth disease virus transmission dynamics and persistence in a herd of vaccinated dairy cattle in India, *Transbound. Emerg. Dis.*, 65(2), e404–e415.

Heilmann, M., Lkhagvasuren, A., Adyasuren, T., Khishgee, B., Bold, B., Ankhanbaatar, U., Fusheng, G., Raizman, E. and Dietze, K. (2020), African swine fever in Mongolia: course of the epidemic and applied control measures, *Vet. Sci.*,7(1), 24.

Hess, W. R., Endris, R. G., Armando, L. and Manuel, C. J. (1989), Clearance of African swine fever virus from infected tick (acari) colonies, *J. Med. Entomol.*, 4, 314–317.

Iglesias, I., Muñoz, M. J., Montes, F., Perez, A., Gogin, A., Kolbasov, D. and de la Torre, A. (2016), Reproductive ratio for the local spread of African swine fever in wild boars in the Russian Federation, *Transbound. Emerg. Dis.*, 63(6), e237–e245.

Jori, F. and Bastos, A. D. (2009), Role of wild suids in the epidemiology of African swine fever, *EcoHealth*, 6(2), 296–310.

Jun-Sik, L., Eutteum, K., Pan-Dong, R. and Son-Il, P. (2021), Basic reproduction number of African swine fever in wild boars (Sus scrofa) and its spatiotemporal heterogeneity in South Korea, *J. Vet. Sci.*, 22(5), 71.

Le, V. P., Jeong, D. G., Yoon, S. W., Kwon, H. M.,Trinh, T. B. N., Nguyen, T. L., Bui, T. T. N., Oh, J., Kim, J. B., Cheong, K. M., Van Tuyen, N., Bae, E., Vu, T. T. H., Yeom, M., Na, W. and Song, D. (2019), Outbreak of African swine fever, Vietnam, 2019, *Emerg. Infect. Dis.*, 25(7), 1433–1435.

Li, J., Jin, Z., Wang, Y., Sun, X., Xu, Q., Kang, J., Huang, B. and Zhu, H. (2021), Data-driven dynamical modelling of the transmission of African swine fever in a few places in China, *Transbound.Emerg. Dis.*, 69(4), e646–e658.

Loi, F., Cappai, S., Laddomada, A., Feliziani, F., Oggiano, A., Franzoni, G., Rolesu, S. and Guberti, V. (2020), Mathematical approach to estimating the main epidemiological parameters of African swine fever in wild boar, *Vaccines*, 8(3), 521.

Lyra, L., Saraiva, V., Lage, G. and Samarcos, M. (1986), Eradication of African swine fever from Brazil. *Revue Scientifique Et Technique De Loie Revue Scientifique et Technique de l'OIE*, 5(3), pp. 771–787.

Lyra, T. M. (2006), The eradication of African swine fever in Brazil, 1978-1984, *RST*, 25(1), 93–103.

McVicar, J. W., Mebus, C. A., Becker, H. N., Belden, R. C. and Gibbs, E. P. (1981), Induced African swine fever in feral pigs, *J. Am. Vet. Med. Assoc.*, 179(5), 441–446.

Mellor, P. S., Kitching, R. P. and Wilkinson, P. J. (1987), Mechanical transmission of capripox virus and African swine fever virus by Stomoxys calcitrans, *Res. Vet. Sci.*, 43(1), 109–112.

Montgomery, R. E. (1921), On aform of swine fever occurring in British East Africa (Kenya Colony), *J. Comp. Pathol. Ther.*, 34, 159–191.

Morilla, A., Yoon, K. J. and Zimmerman, J. J. (2002), Trends in emerging viral infections of swine (Zimmerman/trends), *ASF*, 119–124.

Mulumba-Mfumu, L. K., Saegerman, C., Dixon, L. K., Madimba, K. C., Kazadi, E., Mukalakata, N. T., Oura, C. A. L., Chenais, E., Masembe, C., Ståhl, K., Thiry, E. and Penrith, M. L. (2019), African swine fever: update on Eastern, Central and Southern Africa, *Transbound. Emerg. Dis.*, 66(4), 1462–1480.

Mur, L., Martínez-López, B. and Sánchez-Vizcaíno, J. M. (2012), Risk of African swine fever introduction into the European Union through transport-associated routes: returning trucks and waste from international ships and planes, *BMC Vet. Res.*, 8(1), 149.

Njau, E. P., Domelevo Entfellner, J. B., Machuka, E. M., Bochere, E. N., Cleaveland, S., Shirima, G. M., Kusiluka, L. J., Upton, C., Bishop, R. P., Pelle, R. and Okoth, E. A.

(2021), The first genotype II African swine fever virus isolated in Africa provides insight into the current Eurasian pandemic, *Sci. Rep.*, 11(1), 13081.

OIE (2022), *World Animal Health Information System*. World Organisation for Animal Health Website.

Oie, A. H. S. (2019), *Terrestrial Animal Health Code* (28th edn.) (vol. 2019), France.

Owolodun, O. A., Obishakin, E. T., Ekong, P. S. and Yakubu, B. (2010), Investigation of African swine fever in slaughtered pigs, Plateau state, Nigeria, 2004-2006, *Trop. Anim. Health Prod.*, 42(8), 1605-1610.

Patil, S. S., Suresh, K. P., Vashist, V., Prajapati, A., Pattnaik, B. and Roy, P. (2020), African swine fever: A permanent threat to Indian pigs, *Vet. World*, 13(10), 2275-2285.

Pei, J., Hu, Z., Xie, J., Cao, Q., Wang, G. and Song, N. (2021), Emergency epidemiological investigation on African swine fever in wild boars in Shennongjia forestry district of Hubei Province, *China Anim. Health Inspect.*, 38(1), 7.

Pietschmann, J., Guinat, C., Beer, M., Pronin, V., Tauscher, K., Petrov, A., Keil, G. and Blome, S. (2015), Course and transmission characteristics of oral low-dose infection of domestic pigs and European wild boar with a Caucasian African swine fever virus isolate, *Archi. Virol.*, 160(7), 1657-1667.

Plowright, W., Parker, J. and Peirce, M. A. (1969), African swine fever virus in Ticks (Ornithodoros moubata, Murray) collected from Animal Burrows in Tanzania, *Nature*, 221(5185), 1071-1073.

Quembo, C. J., Jori, F., Vosloo, W. and Heath, L. (2018), Genetic characterization of African swine fever virus isolates from soft ticks at the wildlife/domestic interface in Mozambique and identification of a novel genotype, *Transbound. Emerg. Dis.*, 65(2), 420-431.

Ravaomanana, J., Michaud, V., Jori, F., Andriatsimahavandy, A., Roger, F., Albina, E. and Vial, L. (2010), First detection of African swine fever virus in Ornithodoros porcinus in Madagascar and new insights into tick distribution and taxonomy, *Parasites Vectors*, 3(1), 115-115.

Rennie, L., Wilkinson, P. J. and Mellor, P. S. (2010), Transovarial transmission of African swine fever virus in the argasid tick Ornithodoros moubata, *Med. Vet. Entomol.*, 15(2), 140-146.

Sánchez-Cordón, P. J., Chapman, D., Jabbar, T., Reis, A. L., Goatley, L., Netherton, C. L., Taylor, G., Montoya, M. and Dixon, L. (2017), Different routes and doses influence protection in pigs immunised with the naturally attenuated African swine fever virus isolate OURT88/3, *Antiviral Res.*, 138, 1-8.

Sánchez-Vizcaíno, J. M., Martínez-López, B., Martínez-Avilés, M., Martins, C., Boinas, F., Vialc, L., Michaud, V., Jori, F., Etter, E., Albina, E. and Roger, F. (2009), Scientific review on African Swine Fever, *EFSA*, 6(8).

Sánchez-Vizcaíno, J. M., Mur, L., Gomez-Villamandos, J. C. and Carrasco, L. (2015), An update on the epidemiology and pathology of African swine fever, *J. Comp. Pathol.*, 152(1), 9-21.

Sánchez-Vizcaíno, J. M., Mur, L. and Martınez-Lopez, B. (2012), African swine fever: an epidemiological update, *Transbound. Emerg. Dis.*, 59 (Suppl. 1), 27-35.

Sauter-Louis, C., Conraths, F. J., Probst, C., Blohm, U., Schulz, K., Sehl, J., Fischer, M., Forth, J. H., Zani, L., Depner, K., Mettenleiter, T. C., Beer, M. and Blome, S. (2021), African swine fever in wild boar in Europe—a review, *Viruses*, 13(9), 1717.

Schulz, K., Conraths, F. J., Blome, S., Staubach, C. and Sauter-Louis, C. (2019), African swine fever: fast and furious or slow and steady? *Viruses*, 11(9).

Ster, I. C., Dodd, P. J. and Ferguson, N. M. (2012), Within-farm transmission dynamics of foot and mouth disease as revealed by the 2001 epidemic in Great Britain, *Epidemics*, 4(3), 158-169.

Sun, E., Huang, L., Zhang, X., Zhang, J., Shen, D., Zhang, Z., Wang, Z., Huo, H., Wang, W.,Huangfu, H., Wang, W., Li, F., Liu, R., Sun, J., Tian, Z., Xia, W., Guan, Y., He, X., Zhu, Y., Zhao, D. and Bu, Z. (2021), Genotype I African swine fever viruses emerged in domestic pigs in China and caused chronic infection, *Emerg. Microbes Infect.*, 10(1), 2183-2193.

Thomson, G. R. (1985), The epidemiology of African swine fever: the role of free-living hosts in Africa, *Onderstepoort J. Vet. Res.*, 52(3), 201-209.

Walczak, M., Mudzki, J., Mazur-Panasiuk, N., Juszkiewicz, M. and Woniakowski, G. (2020), Analysis of the clinical course of experimental infection with highly pathogenic African swine fever strain, isolated from an outbreak in Poland. Aspects related to the disease suspicion at the farm level, *Pathogens*, 9(3).

Wang, Y., Kang, W., Yang, W., Zhang, J., Li, D. and Zheng, H. (2021), Structure of African swine fever virus and associated molecular mechanisms underlying infection and immunosuppression: areview, *Front. Immunol.*, 12, 715582.

Wang, N., Zhao, D., Wang, J., Zhang, Y., Wang, M., Gao, Y., Li, F., Wang, J., Bu, Z., Rao, Z. and Wang, X. (2019), Architecture of African swine fever virus and implications for viral assembly, *Science*, 366(6465), 640-644.

Wardley, R. C., Andrade, CdM., Black, D. N., de Castro Portugal, F. L., Enjuanes, L., Hess, W. R., Mebus, C., Ordas, A., Rutili, D., Sanchez Vizcaino, J., Vigario, J. D., Wilkinson, P. J., Moura Nunes, J. F. and Thomson, G. (1983), African swine fever virus, *Arch. Virol.*, 76(2), 73-90.

Yoo, D., Kim, H., Lee, J. Y. and Yoo, H. S. (2020), African swine fever: etiology, epidemiological status in Korea, and perspective on control, *J. Vet. Sci.*,21(2), e38.

Zani, L., Forth, J. H., Forth, L., Nurmoja, I., Leidenberger, S., Henke, J., Carlson, J., Breidenstein, C., Viltrop, A., Hoper, D., Sauter-Louis, C., Beer, M. and Blome, S. (2018), Deletion at the 5'-end of Estonian ASFV strains associated with an attenuated phenotype, *Sci. Rep.*, 8(1), 6510.

Zhang, Y., Zhang, J., Yang, J., Yang, J., Han, X., Mi, L., Zhang, F., Yu, Q., Zhang, S., Wang, Y., Zhou, X., Yue, H. and Wang, S. (2021), Identification of a natural variant of African Swine Fever in China, *J. Vet. Sci.*, 41(2).

Chapter 2

Advances in surveillance and diagnostic techniques for tracking the spread of African swine fever

Sandra Blome, Federal Research Institute for Animal Health – Friedrich Loeffler Institute, Germany

1 Introduction

This chapter begins with a brief overview of established practices relating to the types and preparation of samples for analysis as well as current diagnostic tools. It then goes on to explore three areas that go beyond established diagnostics techniques and reflect recent trends in research:

- The use of alternative sample matrices for African swine fever (ASF) diagnosis in domestic and wild pigs;
- Pen-side diagnostics; and
- Recent advances in genomic epidemiology (especially in areas with the occurrence of local variants).

All these areas are helping to improve the speed, flexibility, and accuracy of disease identification and surveillance.

2 Sample matrices

Diagnosis of notifiable animal diseases is much more than laboratory diagnostics. The most important element in early detection is the person who

http://dx.doi.org/10.19103/9781786768629.02

suspects (or even just wants to rule out) the disease and takes the right samples from the right animals at the right time. Even the best test cannot work reliably without the right sampling scheme and quality of samples. This means sampling schemes and targets have to be properly defined.

When choosing the routine sample matrix, it should be kept in mind that ASF is originally a vector-borne disease optimized for transmission via small amounts of blood and only moderately contagious in the absence of vectors and prominent hemorrhagic signs. Blood contains high viral loads and viremia is persistent, even in convalescent animals. The viral genome and its parts remain detectable for a long period, even after infectivity and the presence of an infectious virus has subsided. Examples from past animal experiments show that the viral genome is detectable as early as 2-4 days after infection and in a high percentage of animals until day 90 or even slightly beyond [1].

While we do find viruses and viral genomes in secretions and excretions, their suitability for the diagnosis of a single animal is limited and sometimes inconsistent [2]. Blood is usually the sample of choice when aiming at early pathogen detection in live animals. When comparing ethylenediamine tetraacetic acid (EDTA) blood and serum as matrix, EDTA blood is superior, especially in the early and late phases of infection [3]. When PCR is considered, it should also be noted that heparin is a strong inhibitor of any PCR reaction.

As a general rule, it should be noted that an animal found dead or visibly ill is always preferable to a random sample, even in small groups. When choosing sample matrices for the detection of African swine fever virus (ASFV) in carcasses, spleen samples are preferable. However, several other organs, including lymph nodes, tonsil, lung, kidney, and bone marrow will give reliable results. There is also no difference in testing domestic and wild suids if basic sample quality can be ensured [3].

Where antibody detection techniques are used to supplement outbreak investigations (particularly in areas or countries with endemic ASFV circulation and vaccination), serum or plasma is usually the sample matrix of choice [4]. However, meat juice, filter paper or swab punches, and oral fluids may also be appropriate [5-8]. First positive results are to be expected from 8 to 15 days post-infection. It is important to note that antibodies persist over long periods of time, i.e. months or even years.

3 Detection methods

Provided that general laboratory facilities are suitable, there is a range of reliable direct and indirect detection methods available for the early detection and surveillance of ASF. Most laboratories use polymerase chain reaction (PCR) for the detection of the viral genome and antibody enzyme-linked immunosorbent assays (ELISAs) as a first line of defense. Confirmatory tests are in place to verify ambiguous results. These methods are described in international manuals [4, 9] and laboratory operating procedures (e.g. the EURL standard operating procedures,   https://asf-referencelab.info/asf/en/procedures-diagnosis/sops) and have been validated in national and international reference laboratories. An overview of routinely available methods is shown in Fig. 1.

Many molecular techniques (real-time PCRs and isothermal amplification methods) and enzyme immunoassays have been commercialized, especially in the last years of the panzootic spread of ASF [3, 10]. Commercial tests can have the advantage of external validation and, in some countries, licensing and batch release. It is always advisable to choose tests or test kits that have been validated by a reference laboratory at an international or national level.

For screening purposes, different ELISA systems are available that utilize different antigens (e.g. p72, p30, p54) and formats (competitive, indirect). Immunoblotting, indirect immunofluorescence, and immunoperoxidase tests are available (see Fig. 1). The latter tests are important to confirm or rule out positive serological results, especially when testing samples of limited quality (e.g. from wild boar).

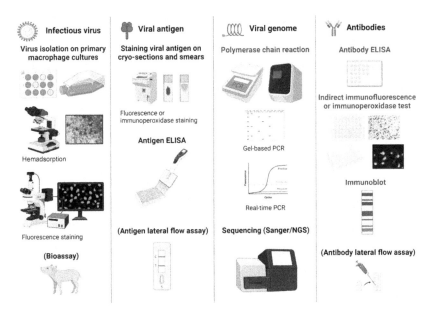

Figure 1 Overview of the routinely available diagnostic techniques for African swine fever. The two techniques often used as the first line of defense, e.g. PCR and antibody ELISA, are highlighted in dark red, confirmatory serological assays are displayed in dark green. Supplementary techniques are displayed in brackets. Source: Created with BioRender.com.

4   Alternative sample matrices

As described previously, there is no doubt that a decent blood or spleen sample is the matrix of choice for pathogen detection in most ASF diagnostic scenarios [3]. However, there are several circumstances where blood/spleen sampling could be difficult/impossible or just much more laborious. Among those situations is carcass-based passive surveillance in domestic and wild pigs, and early warning in large farms.

When considering alternative matrices, we have to keep in mind that non-invasive sample matrices such as oral or fecal swabs contain much less virus/viral genome [1, 2, 11, 12] and may lead to false negative overall results [13]. Thus, alternative matrices have to show fitness for purpose and have to be embedded into diagnostic workflows [3].

One of the challenging diagnostic situations where alternative sample matrices could be considered is sampling of wild boar, especially wild boar carcasses. In many countries in Europe, cases of ASF predominate in wild boar [14]. Here it is not uncommon to need to adapt routine diagnostics, as conservation status and accessibility of samples can be very limited [15].

To find a pragmatic sampling matrix for these conditions, Petrov et al. [16] assessed the use of different dry blood swabs for ASFV diagnosis. The advantage is that the swab is already combined with a shipment-suitable receptacle, and no direct contact or further equipment is needed. For use in real-time PCR, either the swab can be submerged in the lysis buffer of the extraction kit or a small piece of the swab (a diamond-shaped piece) can be cut and used in an organ protocol.

As an example, extraction using the QIAamp Viral RNA Mini Kit (Qiagen; generates nucleic acids that can be used for both classical swine fever virus and ASFV detection) is done as follows:

- A diamond-shaped swab piece is cut from the swab and transferred directly into the lysis buffer of the kit (AVL).
- After thorough vortexing, the sample is incubated for 10 min at room temperature.
- After brief centrifugation, 560 µL ethanol is added and the process is continued according to the manufacturer's instructions.

An exemplary overall workflow is depicted in Fig. 2. In reported studies, so-called Genotubes (ThermoFisher) were the optimum in terms of handling and stability. It has been demonstrated that reliable ASFV genome detection is possible over a wide range of viral loads and that, with small limitations, blood swabs are also suitable for antibody detection [17].

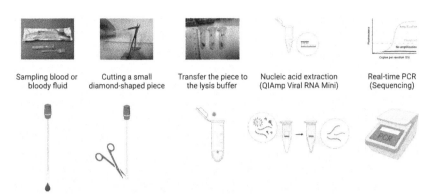

Figure 2 PCR workflows for dry blood swabs. After sampling blood or bloody fluid on the swab, a small diamond-shaped piece is cut and transferred to the lysis buffer of the respective extraction kit. From that point, the extraction is carried out according to the manufacturer's instructions. The obtained nucleic acid is used for downstream analyses (PCR and/or sequencing). Source: Created with BioRender.com.

More recently, these initial data were supplemented with a broader validation study that included the use of antibody lateral flow devices [7]. Using Genotubes, routine qPCR reached a sensitivity of 98.8% (CI 93.4, 100.0) and a specificity of 98.1% (CI 90.1, 100.0) under laboratory conditions [85.7% (CI 71.5, 99.6) with stored field samples]. Serology by ELISA achieved 93.1% sensitivity (CI 83.3, 98.1) and 100% specificity (CI 95.9, 100.0). This approach is particularly interesting as it was shown that it had almost no problems with bad sample quality.

Following the availability of new swabs and transport media (e.g. PrimeStore MTM, Longhorn Vaccines, and Diagnostics), the technique was reassessed using samples from experimental studies [3]. The conserving transport media, in particular, showed high performance and surpassed all dry blood swabs. Under the reported settings, the viral genome loads were always lower on swabs or in transport media, but the qualitative result remained the same. It is noteworthy that the blood swab is now being used routinely (along with other sample matrices) to monitor the current German outbreak of ASF [18].

In general, this swab approach is also suitable for the sampling of fallen domestic animals with suspected ASF. However, in this context, suitable organ samples could be superior. Sample matrices that could be collected without the need of opening the carcass at the point of origin have been evaluated, thus avoiding contamination. Even ear notches or ocular fluids will give a positive result in most animals that have died from ASF [3, 19, 20]. However, easily accessible inguinal lymph nodes performed much better and have been assessed in different laboratories [3, 21]. Goonewardene et al. [21] report that genome copy numbers in superficial inguinal lymph nodes correlated well with those in the spleen and, by sampling those lymph nodes, all pigs could be detected that had succumbed to highly virulent and moderately virulent ASFV strains (100% sensitivity). ASFV was also isolated from all positive lymph node samples. Thus, sampling superficial inguinal lymph nodes could be useful in routine surveillance of dead pigs on commercial and backyard farms, mitigating contamination risks while still ensuring diagnostic quality.

Given the broad range of rather unspecific clinical signs related to ASFV infection [22], early warning is crucial and necessitates laboratory investigations. For routine background screening without concrete suspicion that ASF might be present, non-invasive samples would provide a valuable approach. Pen-based oral fluid collection for active surveillance could be a non-invasive alternative that is less resource and time-intensive than blood collection. The technique has been used for other diseases with great success [23]. Recently, Goonewardene et al. [24] showed in four independent animal experiments with different ASFV strains that ASFV genome could be detected in individual blood samples as early as 1-day post-infection and detected in oral fluids at low-to-moderate levels as early as 3–5 days post-infection in all four independent experiments. The authors conclude that pen-based oral fluid samples may be used to supplement the use of traditional samples for rapid detection of ASFV during ASF surveillance. The same conclusions are drawn by Lee et al. [19] after comparing different matrices upon infection with an ASFV strain from Vietnam. Oral fluids showed also potential for antibody detection [5]. While it may not be the matrix that ensures the highest sensitivity, oro-fluids may still add to our diagnostic toolbox but need to be embedded into established routines.

5   Pen-side diagnostics

In recent years, the idea of pen-side testing has been investigated to enable diagnostics under field conditions and in regions with limited laboratory capacity. On the one hand, different direct immunochromatographic techniques (lateral flow assays – LFA) for the detection of ASFV antigens and antibodies have been developed and, on the other hand, PCR-based or isothermal amplification techniques have been optimized for field use.

The easiest approach is probably LFA. While antibody detection using LFAs is quite sensitive and almost comparable to ELISA [3, 7], much lower sensitivity was observed when detecting viral antigens using LFA [25]. However, field experience showed that the combination of an antigen lateral flow device (LFD) (in this case the INgezim PPA CROM Ag LFD, Ingenasa) used on sick animals and an antibody LFD (INgezim PPA CROM Ab LFD, Ingenasa) can help to prioritize sampling and control efforts in pig farms affected by ASF [26]. The same was concluded for a wild boar/feral pig setting [27]. However, field experience from Germany showed that negative results in antigen LFA must be regarded with great care [25]. Even samples with very high viral loads detected in real-time PCR did not result in positive reactions.

New pen-side tests have been reported and commercialized in recent years. Among them are the PenCheck (Silver Lake Research) and the Eradikit African Swine Fever Lateral Flow Assay (In3diagnostic). The former is suitable for both blood and tissues, the latter mainly for tissues. Both tests have shown potential under experimental conditions and with defined field materials [28; and Friedrichs et al., manuscript in preparation]. Related techniques have been used such as colloidal–gold dual immunochromatography for the detection of p30- and p72-specific antibodies [29] or fluorescent immunochromatography [30]. There is a need to further test field use performance.

In recent years, PCR-based systems have been very successfully transferred to pen-side applications, ready for practical use [31–34]. Several other amplification techniques have been tried but need much more validation before they can be considered as fit-for-purpose.

Isothermal amplification methods can aid diagnosis under basic laboratory or even field settings and have been investigated recently (see GARA gap analysis report, https://www.ars.usda.gov/GARA/). Among the isothermal methods that have been tried for ASFV is cross-priming amplification (targeting the p72 gene). Fraczyk et al. [35] showed that the method met WOAH–sensitivity recommendations for UPL PCR, at least under selected conditions. Similar results were obtained for a loop-mediated isothermal amplification (LAMP) assay targeting the topoisomerase II gene. Apart from standard settings, visualization through the use of dual-labeled biotin and fluorescein amplicons on lateral flow devices has been demonstrated [36]. Promising LAMP assays have also been reviewed targeting the p10 gene of ASFV [37] and a portable LAMP assay coupled with a clustered regularly interspaced short palindromic repeats (CRISPR)/Cas12a system [38]. Other LAMP applications were, e.g. combined with Lateral Flow Dipsticks [39].

Apart from other isothermal methods such as recombinase polymerase amplification (RPA) (as implemented, e.g. by TwistDx or recombinase-aided amplification developed by Qitian [40] or Wu et al. [41]), different point-of-care isothermal methods have been tested in recent years. Among them are a combination of RPA with a gold nanoparticle test strip [42], RPA-CRISPR-based assays [43] and a field-deployable C-SAND assay kit [44]. Digital PCR and CRISPR-Cas-related assays have also been explored. An example of the latter is a point-of-care/pen-side system that employs a CRISPR-Cas12a and fluorescence system [45]. Another recent approach was the design of ladder-shaped melting temperature isothermal amplification (LMTIA) assays [46] or graphene oxide-based accelerated strand exchange amplification (GO-ASEA) [47]. However, all these methods need further assessment when it comes to field application.

6   Developments in genomic epidemiology

Over the last 10–15 years, whole genome sequences of all relevant ASFV genotypes have been completed as well as sequencing of strains involved in the current ASF outbreak. Numerous sequences have been generated to identify genetic markers to aid genomic epidemiology, understand virulence, and design tailored diagnostics and vaccines [48, 49].

However, these initiatives have lacked both quality and quantity [48]. Unfortunately, next-generation sequencing of ASFV is quite demanding due to extensive homopolymer and repeat regions and the ratio of the host to virus is often very poor when sequencing directly from a diagnostic specimen [49]. To overcome these problems, specific target enrichment (e.g. with probe libraries) and host genome reduction have been tried both separately and in combination [49, 50]. Moreover, new sequencing platforms have been used or combined to establish more reliable workflows and datasets [49, 51]. There is still a need for more work on the quality and harmonization of sequencing approaches, annotation of viral genes, harmonization, and reporting of quality parameters. To provide a platform with accurate and up-to-date information, the Center of Excellence for African Swine Fever Genomics (ASFV Genomics) has been recently launched and provides opportunities to share findings.

Apart from technical issues, genomic epidemiology of ASFV is hampered by the fact that the virus can be rather stable over both space and time. When looking at genotype II strains from ten affected countries, only two intergenic regions have been identified that show significant differences through tandem repeats [52]. While closely related strains may be traceable [52, 53], none of those regions have proven useful for genomic epidemiology at a larger scale [48].

However, the recent discovery of distinct ASFV lineages and variants at the border between Germany and Poland underlines the importance of generating whole ASFV genomes. Five clearly distinguishable lineages and more than ten different variants have been identified when sequencing representative strains originating from different regions in Germany that had ASFV introductions from September 2020 [54]. Variants are characterized by insertions/deletions in homopolymer and non-homopolymer regions as well as synonymous, missense, and nonsense mutations found in annotated genes and intergenomic regions (see Fig. 3). Interestingly, all new variants share a frameshift mutation in the 3′ end of the DNA polymerase PolX gene O174L. This mutation had already been reported from Poland [55] and could be the driver for accelerated evolution.

Using tailored Sanger sequencing approaches, clear spatial and temporal clustering of variants has been shown and meaningful genomic epidemiology has become possible for the first time [54]. Variants of these lineages keep evolving and, to make the most of the data set, mutation rate and evolutionary

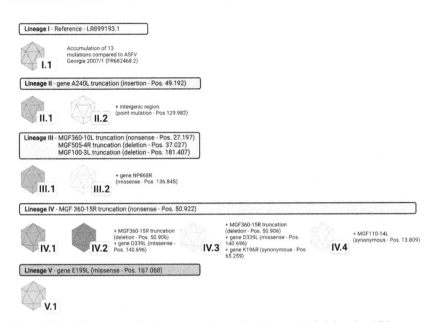

Figure 3 Viral lineages and variants occurring at the German–Polish border. All lineages share a frameshift mutation in the O174L gene and show changes in the core genome and the multigene families. Source: Modified from: Forth et al. (2022).

hotspots should be addressed in the near future. It could be shown, for the German context, that genomic epidemiology can provide insights into virus origin and spread but, even more importantly, help assess the success of control measures in wild boar (where some variants become extinct while others thrive).

To ease the workflow for genomic epidemiology, mutations have now been used to establish multiplex real-time PCRs for the main German lineages along with ASF detection (Deutschmann et al., manuscript submitted). To achieve specific detection of the single nucleotide polymorphisms (SNP), locked nucleic acid (LNA) probes were used for the assays. In general, C-SAND assays targeting the SNPs (MatMaCorp) were also able to distinguish those lineages (Deutschmann et al., manuscript in preparation). This German example could be a blueprint for other regions where distinct variants are detected.

Another application of genomic epidemiology and the use of tailored diagnostics is the tracing of viral genotypes and strains in regions with (legal and illegal) vaccination against ASF using live vaccines. For those purposes, real-time PCRs and other genome detection methods targeting the deleted part of the genome or other strain-specific sequences can be employed that were designed along with the vaccines [56] or later in the country dealing with the challenge [57, 58-62].

7 Conclusion

In conclusion, routine diagnostic techniques have been developed following the panzootic spread of ASFV. The availability of sensitive detection methods allows for pragmatic and alternative sampling, promising pen-side approaches have been developed, and genomic epidemiology is gaining importance in areas with the occurrence of different ASFV lineages and variants.

8 References

1. Petrov, A., Forth, J. H., Zani, L., Beer, M. and Blome, S. No evidence for long-term carrier status of pigs after African swine fever virus infection. *Transbound. Emerg. Dis.* 2018 65(5): 1318–1328.
2. de Carvalho Ferreira, H. C., Weesendorp, E., Quak, S., Stegeman, J. A. and Loeffen, W. L. Suitability of faeces and tissue samples as a basis for non-invasive sampling for African swine fever in wild boar. *Vet. Microbiol.* 2014 172(3-4): 449-454.
3. Pikalo, J., Deutschmann, P., Fischer, M., Roszyk, H., Beer, M. and Blome, S. African swine fever laboratory diagnosis-lessons learned from recent animal trials. *Pathogens* 2021 10(2).
4. WOAH, African swine fever (infection with African swine fever virus). In: *Manual of Diagnostic Tests and Vaccines for Terrestrial Animals*, W.O.f.A. Health, Editor 2022.
5. Mur, L., Gallardo, C., Soler, A., Zimmermman, J., Pelayo, V., Nieto, R., Sánchez-Vizcaíno, J. M. and Arias, M. Potential use of oral fluid samples for serological diagnosis of African swine fever. *Vet. Microbiol.* 2013 165(1-2): 135-139.
6. Randriamparany, T., Kouakou, K. V., Michaud, V., Fernández-Pinero, J., Gallardo, C., Le Potier, M. F., Rabenarivahiny, R., Couacy-Hymann, E., Raherimandimby, M. and Albina, E. African swine fever diagnosis adapted to tropical conditions by the use of dried-blood filter papers. *Transbound. Emerg. Dis.* 2016 63(4): 379-388.
7. Carlson, J., Zani, L., Schwaiger, T., Nurmoja, I., Viltrop, A., Vilem, A., Beer, M. and Blome, S. Simplifying sampling for African swine fever surveillance: assessment of antibody and pathogen detection from blood swabs. *Transbound. Emerg. Dis.* 2018 65(1): e165–e172.
8. Onyilagha, C., Nash, M., Perez, O., Goolia, M., Clavijo, A., Richt, J. A. and Ambagala, A. Meat exudate for detection of African swine fever virus genomic material and anti-ASFV antibodies. *Viruses* 2021 13(9): 1744.
9. Beltran-Alcrudo, D., Arias, M., Gallardo, C., Kramer, S. A., Penrith, M.-L., Kamata, A. and Wiersma, L. *African Swine Fever: Detection and Diagnosis* 2017: Food and agriculture Organization of the United Nations (FAO).
10. Pikalo, J., Carrau, T., Deutschmann, P., Fischer, M., Schlottau, K., Beer, M. and Blome, S. Performance characteristics of Real-time PCRs for African swine fever virus genome detection-comparison of twelve kits to an OIE-recommended method. *Viruses* 2022 14(2): 220.
11. Pietschmann, J., Guinat, C., Beer, M., Pronin, V., Tauscher, K., Petrov, A., Keil, G. and Blome, S. Course and transmission characteristics of oral low-dose infection of domestic pigs and European wild boar with a Caucasian African swine fever virus isolate. *Arch. Virol.* 2015 160(7): 1657-1667.

12. Guinat, C., Reis, A. L., Netherton, C. L., Goatley, L., Pfeiffer, D. U. and Dixon, L. Dynamics of African swine fever virus shedding and excretion in domestic pigs infected by intramuscular inoculation and contact transmission. *Vet. Res.* 2014 45(1): 93.

13. Walczak, M., Szczotka-Bochniarz, A., Żmudzki, J., Juszkiewicz, M., Szymankiewicz, K., Niemczuk, K., Pérez-Núñez, D., Liu, L. and Revilla, Y. Non-invasive sampling in the aspect of African swine fever detection – A risk to accurate diagnosis. *Viruses* 2022 14(8): 1756.

14. Chenais, E., Depner, K., Guberti, V., Dietze, K., Viltrop, A. and Ståhl, K. Epidemiological considerations on African swine fever in Europe 2014-2018. *Porcine Health Manag.* 2019 5: 6.

15. Carlson, J., Zani, L., Schwaiger, T., Nurmoja, I., Viltrop, A., Vilem, A., Beer, M. and Blome, S. Simplifying sampling for African swine fever surveillance: assessment of antibody and pathogen detection from blood swabs. *Transbound. Emerg. Dis.* 2018 65(1): e165–e172.

16. Petrov, A., Schotte, U., Pietschmann, J., Dräger, C., Beer, M., Anheyer-Behmenburg, H., Goller, K. V. and Blome, S. Alternative sampling strategies for passive classical and African swine fever surveillance in wild boar. *Vet. Microbiol.* 2014 173(3–4): 360-365.

17. Blome, S., Goller, K. V., Petrov, A., Dräger, C., Pietschmann, J. and Beer, M. Alternative sampling strategies for passive classical and African swine fever surveillance in wild boar–Extension towards African swine fever virus antibody detection. *Vet. Microbiol.* 2014 174(3–4): 607-608.

18. Sauter-Louis, C., Forth, J. H., Probst, C., Staubach, C., Hlinak, A., Rudovsky, A., Holland, D., Schlieben, P., Göldner, M., Schatz, J., Bock, S., Fischer, M., Schulz, K., Homeier-Bachmann, T., Plagemann, R., Klaaß, U., Marquart, R., Mettenleiter, T. C., Beer, M., Conraths, F. J. and Blome, S. Joining the club: first detection of African swine fever in wild boar in Germany. *Transbound. Emerg. Dis.* 2021 68(4): 1744-1752.

19. Lee, H. S., Bui, V. N., Dao, D. T., Bui, N. A., Le, T. D., Kieu, M. A., Nguyen, Q. H., Tran, L. H., Roh, J. H., So, K. M., Hur, T. Y. and Oh, S. I. Pathogenicity of an African swine fever virus strain isolated in Vietnam and alternative diagnostic specimens for early detection of viral infection. *Porcine Health Manag.* 2021 7(1): 36.

20. Flannery, J., Ashby, M., Moore, R., Wells, S., Rajko-Nenow, P., Netherton, C. L. and Batten, C. Identification of novel testing matrices for African swine fever surveillance. *J. Vet. Diagn. Invest.* 2020 32(6): 961-963.

21. Goonewardene, K. B., Onyilagha, C., Goolia, M., Le, V. P., Blome, S. and Ambagala, A. Superficial inguinal lymph nodes for screening dead pigs for African swine fever. *Viruses* 2022 14(1): 83.

22. Gallardo, C., Nurmoja, I., Soler, A., Delicado, V., Simón, A., Martin, E., Perez, C., Nieto, R. and Arias, M. Evolution in Europe of African swine fever genotype II viruses from highly to moderately virulent. *Vet. Microbiol.* 2018 219: 70-79.

23. Prickett, J. R. and Zimmerman, J. J. The development of oral fluid-based diagnostics and applications in veterinary medicine. *Anim. Health. Res. Rev.* 2010 11(2): 207-216.

24. Goonewardene, K. B., Chung, C. J., Goolia, M., Blakemore, L., Fabian, A., Mohamed, F., Nfon, C., Clavijo, A., Dodd, K. A. and Ambagala, A. Evaluation of oral fluid as an aggregate sample for early detection of African swine fever virus using four

independent pen-based experimental studies. *Transbound. Emerg. Dis.* 2021 68(5): 2867-2877.

25. Deutschmann, P., Pikalo, J., Beer, M. and Blome, S. Lateral flow assays for the detection of African swine fever virus antigen are not fit for field diagnosis of wild boar carcasses. *Transbound. Emerg. Dis.* 2022 69(4): 2344-2348.

26. Lamberga, K., Depner, K., Zani, L., Oļševskis, E., Seržants, M., Ansonska, S., Šteingolde, Ž., Bērziņš, A., Viltrop, A., Blome, S. and Globig, A. A practical guide for strategic and efficient sampling in African swine fever-affected pig farms. *Transbound. Emerg. Dis.* 2022 69(5): e2408-e2417.

27. Cappai, S., Loi, F., Coccollone, A., Cocco, M., Falconi, C., Dettori, G., Feliziani, F., Sanna, M. L., Oggiano, A. and Rolesu, S. Evaluation of a commercial field test to detect African swine fever. *J. Wildl. Dis.* 2017 53(3): 602-606.

28. Onyilagha, C., Nguyen, K., Luka, P. D., Hussaini, U., Adedeji, A., Odoom, T. and Ambagala, A. Evaluation of a lateral flow assay for rapid detection of African swine fever virus in multiple sample types. *Pathogens* 2022 11(2): 138.

29. Wan, Y., Shi, Z., Peng, G., Wang, L., Luo, J., Ru, Y., Zhou, G., Ma, Y., Song, R., Yang, B., Cao, L., Tian, H. and Zheng, H. Development and application of a colloidal-gold dual immunochromatography strip for detecting African swine fever virus antibodies. *Appl. Microbiol. Biotechnol.* 2022 106(2): 799-810.

30. Li, C., He, X., Yang, Y., Gong, W., Huang, K., Zhang, Y., Yang, Y., Sun, X., Ren, W., Zhang, Q., Wu, X., Zou, Z. and Jin, M. Rapid and visual detection of African swine fever virus antibody by using fluorescent immunochromatography test strip. *Talanta* 2020 219: 121284.

31. Liu, L., Luo, Y., Accensi, F., Ganges, L., Rodríguez, F., Shan, H., Ståhl, K., Qiu, H. J. and Belák, S. Pre-clinical evaluation of a real-time PCR assay on a portable instrument as a possible field diagnostic tool: experiences from the testing of clinical samples for African and classical swine fever viruses. *Transbound. Emerg. Dis.* 2017 64(5): e31-e35.

32. Elnagar, A., Blome, S., Beer, M. and Hoffmann, B. Point-of-care testing for sensitive detection of the African swine fever virus genome. *Viruses* 2022 14(12).

33. Daigle, J., Onyilagha, C., Truong, T., Le, V. P., Nga, B. T. T., Nguyen, T. L., Clavijo, A. and Ambagala, A. Rapid and highly sensitive portable detection of African swine fever virus. *Transbound. Emerg. Dis.* 2021 68(2): 952-959.

34. Liu, L., Atim, S., LeBlanc, N., Rauh, R., Esau, M., Chenais, E., Mwebe, R., Nelson, W. M., Masembe, C., Nantima, N., Ayebazibwe, C. and Ståhl, K. Overcoming the challenges of pen-side molecular diagnosis of African swine fever to support outbreak investigations under field conditions. *Transbound. Emerg. Dis.* 2019 66(2): 908-914.

35. Fraczyk, M., Woźniakowski, G., Kowalczyk, A., Niemczuk, K. and Pejsak, Z. Development of cross-priming amplification for direct detection of the African swine fever virus, in pig and wild boar blood and sera samples. *Lett. Appl. Microbiol.* 2016 62(5): 386-391.

36. James, H. E., Ebert, K., McGonigle, R., Reid, S. M., Boonham, N., Tomlinson, J. A., Hutchings, G. H., Denyer, M., Oura, C. A., Dukes, J. P. and King, D. P. Detection of African swine fever virus by loop-mediated isothermal amplification. *J. Virol. Methods* 2010 164(1-2): 68-74.

37. Wang, D., Yu, J., Wang, Y., Zhang, M., Li, P., Liu, M. and Liu, Y. Development of a real-time loop-mediated isothermal amplification (LAMP) assay and visual LAMP assay for detection of African swine fever virus (ASFV). *J. Virol. Methods* 2020 276: 113775.

38. Yang, B., Shi, Z., Ma, Y., Wang, L., Cao, L., Luo, J., Wan, Y., Song, R., Yan, Y., Yuan, K., Tian, H. and Zheng, H. LAMP assay coupled with CRISPR/Cas12a system for portable detection of African swine fever virus. *Transbound. Emerg. Dis.* 2022 69(4): e216–e223.

39. Zuo, L., Song, Z., Zhang, Y., Zhai, X., Zhai, Y., Mei, X., Yang, X. and Wang, H. Loop-mediated isothermal amplification combined with lateral flow dipstick for on-site diagnosis of African swine fever virus. *Virol. Sin.* 2021 36(2): 325–328.

40. Fan, X., Li, L., Zhao, Y., Liu, Y., Liu, C., Wang, Q., Dong, Y., Wang, S., Chi, T., Song, F., Sun, C., Wang, Y., Ha, D., Zhao, Y., Bao, J., Wu, X. and Wang, Z. Clinical validation of two recombinase-based isothermal amplification assays (RPA/RAA) for the rapid detection of African swine fever virus. *Front. Microbiol.* 2020 11: 1696.

41. Wu, Y., Yang, Y., Ru, Y., Qin, X., Li, M., Zhang, Z., Zhang, R., Li, Y., Zhang, Z. and Li, Y. The development of a real-time recombinase-aid amplification assay for rapid detection of African swine fever virus. *Front. Microbiol.* 2022 13: 846770.

42. Wang, Z., Yu, W., Xie, R., Yang, S. and Chen, A. A strip of lateral flow gene assay using gold nanoparticles for point-of-care diagnosis of African swine fever virus in limited environment. *Anal. Bioanal. Chem.* 2021 413(18): 4665–4672.

43. Ren, M., Mei, H., Zhou, M., Fu, Z. F., Han, H., Bi, D., Peng, F. and Zhao, L. Development of A super-sensitive diagnostic method for African swine fever using CRISPR techniques. *Virol. Sin.* 2021 36(2): 220–230.

44. Zurita, M., Martignette, L., Barrera, J., Carrie, M., Piscatelli, H., Hangman, A., Brake, D., Neilan, J., Petrik, D. and Puckette, M. Detection of African swine fever virus utilizing the portable MatMaCorp ASF detection system. *Transbound. Emerg. Dis.* 2022 69(5): 2600–2608.

45. He, Q., Yu, D., Bao, M., Korensky, G., Chen, J., Shin, M., Kim, J., Park, M., Qin, P. and Du, K. High-throughput and all-solution phase African swine fever virus (ASFV) detection using CRISPR-Cas12a and fluorescence based point-of-care system. *Biosens. Bioelectron.* 2020 154: 112068.

46. Wang, Y., Wang, B., Xu, D., Zhang, M., Zhang, X. and Wang, D. Development of a ladder-shape melting temperature isothermal amplification (LMTIA) assay for detection of African swine fever virus (ASFV). *J. Vet. Sci.* 2022 23(4): e51.

47. Zhuang, L., Yang, J., Song, C., Sun, L., Zhao, B., Shen, Q., Ren, X., Shi, H., Zhang, Y. and Zhu, M. Accurate, rapid and highly sensitive detection of African swine fever virus via graphene oxide-based accelerated strand exchange amplification. *Anal. Methods* 2022 14(21): 2072–2082.

48. Forth, J. H., Forth, L. F., Blome, S., Höper, D. and Beer, M. African swine fever whole-genome sequencing-Quantity wanted but quality needed. *PLoS Pathog.* 2020 16(8): e1008779.

49. Forth, J. H., Forth, L. F., King, J., Groza, O., Hübner, A., Olesen, A. S., Höper, D., Dixon, L. K., Netherton, C. L., Rasmussen, T. B., Blome, S., Pohlmann, A. and Beer, M. A deep-sequencing workflow for the fast and efficient generation of high-quality African swine fever virus whole-genome sequences. *Viruses* 2019 11(9).

50. Masembe, C., Sreenu, V. B., Da Silva Filipe, A., Wilkie, G. S., Ogweng, P., Mayega, F. J., Muwanika, V. B., Biek, R., Palmarini, M. and Davison, A. J. Genome sequences of five African swine fever virus genotype IX isolates from domestic pigs in Uganda. *Microbiol. Resour. Announc.* 2018 7(13): e01018-18.

51. Granberg, F., Torresi, C., Oggiano, A., Malmberg, M., Iscaro, C., De Mia, G. M. and Belák, S. Complete genome sequence of an African swine fever virus isolate from Sardinia, Italy. *Genome Announc.* 2016 4(6): e01220-16.

52. Gallardo, C., Fernández-Pinero, J., Pelayo, V., Gazaev, I., Markowska-Daniel, I., Pridotkas, G., Nieto, R., Fernández-Pacheco, P., Bokhan, S., Nevolko, O., Drozhzhe, Z., Pérez, C., Soler, A., Kolvasov, D. and Arias, M. Genetic variation among African swine fever genotype II viruses, Eastern and Central Europe. *Emerg. Infect. Dis.* 2014 20(9): 1544–1547.

53. Goller, K. V., Malogolovkin, A. S., Katorkin, S., Kolbasov, D., Titov, I., Höper, D., Beer, M., Keil, G. M., Portugal, R. and Blome, S. Tandem repeat insertion in African swine fever virus, Russia, 2012. *Emerg. Infect. Dis.* 2015 21(4): 731–732.

54. Forth, J. H., Calvelage, S., Fischer, M., Hellert, J., Sehl-Ewert, J., Roszyk, H., Deutschmann, P., Reichold, A., Lange, M., Thulke, H. H., Sauter-Louis, C., Höper, D., Mandyhra, S., Sapachova, M., Beer, M. and Blome, S. African swine fever virus - variants on the rise. *Emerg. Microbes Infect.* 2023 12(1): 2146537.

55. Mazur-Panasiuk, N. and Woźniakowski, G. The unique genetic variation within the O174L gene of Polish strains of African swine fever virus facilitates tracking virus origin. *Arch. Virol.* 2019 164(6): 1667–1672.

56. O'Donnell, V., Holinka, L. G., Gladue, D. P., Sanford, B., Krug, P. W., Lu, X., Arzt, J., Reese, B., Carrillo, C., Risatti, G. R. and Borca, M. V. African swine fever virus Georgia isolate harboring deletions of MGF360 and MGF505 genes is attenuated in Swine and confers protection against challenge with virulent parental virus. *J. Virol.* 2015 89(11): 6048–6056.

57. Guo, Z., Li, K., Qiao, S., Chen, X. X., Deng, R. and Zhang, G. Development and evaluation of duplex TaqMan real-time PCR assay for detection and differentiation of wide-type and MGF505-2R gene-deleted African swine fever viruses. *BMC Vet. Res.* 2020 16(1): 428.

58. Huang, Z., Xu, Z., Cao, H., Zeng, F., Wang, H., Gong, L., Zhang, S., Cao, S., Zhang, G. and Zheng, Z. A triplex PCR method for distinguishing the wild-type African swine fever virus from the deletion strains by detecting the gene insertion. *Front. Vet. Sci.* 2022 9: 921907.

59. Lin, Y., Cao, C., Shi, W., Huang, C., Zeng, S., Sun, J., Wu, J. and Hua, Q. Development of a triplex real-time PCR assay for detection and differentiation of gene-deleted and wild-type African swine fever virus. *J. Virol. Methods* 2020 280: 113875.

60. Li, X., Hu, Y., Liu, P., Zhu, Z., Liu, P., Chen, C. and Wu, X. Development and application of a duplex real-time PCR assay for differentiation of genotypes I and II African swine fever viruses. *Transbound. Emerg. Dis.* 2022 69(5): 2971–2979.

61. Yang, H., Peng, Z., Song, W., Zhang, C., Fan, J., Chen, H., Hua, L., Pei, J., Tang, X., Chen, H. and Wu, B. A triplex real-time PCR method to detect African swine fever virus gene-deleted and wild type strains. *Front. Vet. Sci.* 2022 9: 943099.

62. Zhu, J., Jian, W., Huang, Y., Gao, Q., Gao, F., Chen, H., Zhang, G., Liao, M. and Qi, W. Development and application of a duplex droplet digital polymerase chain reaction assay for detection and differentiation of EP402R-deleted and wild-type African swine fever virus. *Front. Vet. Sci.* 2022 9: 905706.

Chapter 3

Risk-based measures for prevention and control of African swine fever (ASF) in pigs

Silvia Bellini, Istituto Zooprofilattico Sperimentale della Lombardia ed Emilia-Romagna (IZSLER), Italy

1   Introduction

African swine fever virus (ASFV) is a large, enveloped DNA virus classified as the only member of the Asfaviridae family (Dixon et al., 2020); it affects animals belonging to the Suidae family, especially domestic pigs, wild boars and warthogs. The ASF genotype currently circulating in Europe and Asia is a highly virulent and lethal strain (Gogin et al., 2013; Blome et al., 2020). Transmission of ASFV occurs through (Guinat et al., 2016)

- direct contact with infected animals (wild or domestic pigs);
- ingestion of contaminated materials (e.g. swill feed);
- indirect contact with fomites or contaminated objects (bedding, vehicles, equipment, clothes, footwear, etc.); and
- soft tick bites.

An outbreak of ASF in Spain and Portugal (1960s–1980s) involved the presence of the tick *Ornithodors erraticus* in traditionally constructed pig housing, for example, in cracks in adobe or between stones used to construct walls (Wilson et al., 2013). The involvement of ticks in the outbreak in Eastern Europe in 2020 has not been established (Blome et al., 2020). Wild boars are susceptible to ASF,

http://dx.doi.org/10.19103/9781786768629.03

and, where the disease is endemic in wild boar populations (e.g. Europe), wild pigs represent a serious disease threat for domestic pigs (Guberti et al., 2016).

ASFV is able to survive for long periods in a protein-rich environment and remains stable at pH 4–10 (Geering et al., 2001). The ability of ASFV to survive in many environments can play a significant role in the local persistence and geographical spread of the virus (Juszkiewicz et al., 2019), which, over the years, has demonstrated its ability to survive for long enough to cover long distances and infect pigs across continents (Costard et al., 2009). There is evidence of human-mediated transportation of ASFV in some affected countries, and this appears to be an important pathway contributing to the spread of the disease, including the introduction of ASF into Belgium, the Czech Republic, and western Poland reported by European Food Safety Authority (EFSA) (Anon, 2019d).

Concern about ASF reflects the fact that the pig sector is one of the most economically significant farming sectors in the European Union (EU); among terrestrial animals, pork is the most consumed meat, followed by chicken and beef (Anon, 2019a) (Fig. 1). For more than 20 years, ASF was endemic only in the African continent. The unexpected introduction of the ASFV genotype II into the Caucasus in 2007 resulted in the unprecedented geographical spread of the disease. The number of countries or territories reporting the presence of the disease has increased in the past few years, and ASF has officially been notified to the World Organization for Animal Health (OIE) by member countries from sub-Saharan Africa, Europe Asia and Carribean island Hispaniola (OIE-WHAIS, 2021). The spread of ASF is causing serious economic damage in affected

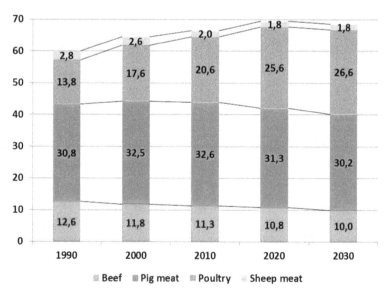

Figure 1 European Union meat consumption by livestock meat type (kg per capita) (Anon, 2019a).

countries, reshaping the pig farming sector and commercial pork production and trade around the world (Bellini, 2021). The occurrence of ASF in Europe has particularly affected smallholder pig holdings, and the control in rural areas with backyard pig production has proven to be difficult.

One of the major obstacles to ASF control is the absence of effective vaccines or treatment to provide adequate protection for both domestic and wild pigs. This means that prevention and early detection play a pivotal role in the control of ASF (Pfeiffer et al., 2021). The EU strategy for the management of ASF follows a risk-based approach with activities focused on ASF prevention and early detection (Anon, 2015). This chapter highlights the surveillance tools currently available for early detection of the presence of ASF and biosecurity measures to prevent its spread, including adequate cleaning and disinfection (C&D) procedures. Farm biosecurity and good farming practice are considered the most effective tools for preventing ASF introduction into pig holdings.

2 Surveillance measures

Animal health surveillance is a key element of disease control policy (Fig. 2). Surveillance is a tool to (Anon, 2019b):

- monitor disease trends;
- identify the presence of the disease at an early stage;
- facilitate the control of infection;
- provide data for use in risk analysis;

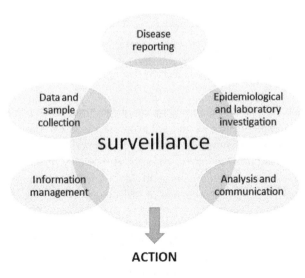

Figure 2 Essential components of a surveillance system (Anon, 2014c).

- substantiate the rationale for the adoption of sanitary measures; and
- provide assurance to trading partners.

In particular, surveillance should enable early detection of the disease so that the relevant authority can put in place early measures for disease prevention and control. Effective implementation of a surveillance system requires a clear rationale and procedures to investigate sources of infection and prevent its spread (Anon, 2014c). Surveillance should take into account the nature of the disease in a particular location. The epidemiology of ASF varies in different regions of the world. Surveillance strategies should be adapted to the local situation, and the sensitivity of the system should be adjusted to the corresponding level of risk (Bellini et al., 2016).

In territory free of the disease, surveillance should aim at early detection of the presence of ASFV through rapid investigation of cases of illness and timely laboratory diagnosis to allow a prompt response and the adoption of proper disease control measures (Anon, 2019b). The spread of ASFV into a previously infected area often occurs through the introduction of contaminated pig products (swill) or new animals into pig farms, while the spread of the disease is often due to poor farming practices (Mur et al., 2016; Plavšic et al., 2019). An important source of data for surveillance is clinical observations of animals in the field. A country's 'early warning system,' which is essential for the timely detection and reporting of diseases, should consider the involvement of stakeholders and promote prompt notification of suspect cases by people working directly in contact with animals (Anon, 2019b).

Clinical surveillance is considered the most effective tool for early detection (Anon, 2019b), with activities focused on:

- identification of clinical signs consistent with ASF; and
- assessment of production and mortality trends on a farm.

This approach is based on detecting variations from the normal range or pattern of key indicators as a warning of potential disease in the herd.

Information management systems in commercial pig holdings routinely track key health and production indices such as mortality, morbidity, veterinary treatments, feed and water consumption. Many of these parameters may be used to detect the potential presence of ASF incursion. They can be grouped as follows:

- Very early indicators (pre-clinical);
  - Data on feed or water consumption; and
  - Data on pig activity level or body temperature.

- Early indicators (clinical);
  - Sudden death; and
  - Health and behavioural observations recorded by barn staff.
- Delayed indicators;
  - Barn mortality records.
- Late indicators;
  - Veterinary diagnosis; and
  - Laboratory tests.

The current value of each indicator (or combination of indicators) should be benchmarked against historical farm data, taking other available factors into account (e.g. level of other diseases, seasonal variation), in order to assess whether the current indicator is within the expected range. If not, an investigation shall be performed to confirm or rule out the presence of ASFV (Pfeiffer et al., 2021). Early detection of ASF on large commercial farms can be more challenging than on small farms since, at the beginning of the epidemic, mortality can be masked by baseline farm mortality. However, a recent study has used mortality data for ASF early detection not only in small pig farms but also in large commercial barns (Faverjon et al., 2020).

For effective early detection, surveillance activities should target pig farms or holdings where the infection is more likely to be introduced or more likely to spread. This includes proximity to high-risk wild and feral populations, pigs reared outdoors (with greater risk of contact with wild populations) and farms that feed swill (Anon, 2019b). Risk factors to be taken into consideration to target surveillance actions may also include (Anon, 2019b):

- areas in which ASF outbreaks have been detected in the past;
- areas at risk due to the prevalence of ASF in adjacent regions or countries; and
- areas where there is evidence of the involvement of ticks in the spread of the disease.

Veterinary authorities should consider these factors when planning risk-based surveillance and control activities. They should also take into account the characteristics of the local pig production system (Bellini et al., 2020). The higher the density of pig farms and susceptible animals, and the higher the rate of indirect or direct contact between pigs and farms, the faster an infectious disease will spread through a population (Dixon et al., 2020).

The criteria used to define a 'suspected case' are key to the sensitivity and specificity of clinical observation. In the EU, a pig holding is defined as 'a suspected holding' based on two main groups of symptoms and pathological findings:

- fever with morbidity and mortality in pigs of all ages; and
- fever with haemorrhagic syndrome, petechial and ecchymotic haemorrhages, especially in the lymph nodes, kidneys, spleen (which is enlarged and dark, particularly in the acute forms) and urinary bladder and ulcerations on the gall bladder.

However, due to the similarity of AFS with other diseases of pigs, clinical surveillance should be supplemented, as appropriate, by serological and virological surveillance (Anon, 2019e).

Virological surveillance is important for early detection and differential diagnosis of ASF. It is used to investigate clinically suspect cases, monitor at-risk populations, monitor sentinel animals, investigate suspected cases or increased mortality, follow up positive serological results and confirm eradication after a stamping-out policy has been applied to eradicate the disease (Anon, 2019b). Serology is also an effective surveillance tool. However, serological surveillance detects the presence of antibodies against ASFV. This means positive ASFV antibody test results indicate an ongoing or a past outbreak, showing that ASF serology is not suitable for early detection (Anon, 2019b).

In the event of suspicion of ASF, the competent authority should immediately conduct an epidemiological investigation to identify the source of the infection in order to control and prevent the spread of the disease. An epidemiological investigation aims to (Anon, 2016):

- identify the likely origin of the disease and the means of transmission;
- calculate the likely length of time that the disease has been present;
- identify farms or other locations where animals may have become infected;
- obtain information on movements of animals, persons, products, vehicles, materials or other means by which the disease agent could have been spread during the period preceding the notification of suspicion or confirmation of the disease; and
- obtain information on the likely spread of the disease in the surrounding environment, including the presence of disease vectors.

Whilst conducting an investigation, the competent authority should put in place preliminary disease control measures to prevent the possible spread of the disease and support its eradication (Anon, 2020). According to EU law, wherever ASF is suspected or confirmed, a combination of actions must be taken to eliminate the source of the pathogen and lead to its eradication:

- slaughter of animals infected or suspected of being infected and safe disposal of dead animals and potentially contaminated products; and
- C&D of affected premises, vehicles and equipment.

In order to mitigate the risk of spread, these measures should be combined with strict preventive measures on farms located in restricted area (Bellini et al., 2016; Jurado et al., 2018). The competent authority may also need to extend disease control measures to other establishments (e.g. processing facilities) which may be implicated in the spread of the disease (Anon, 2020).

Disease control is critically dependent on early detection and an effective chain of disease notification and reporting (Fig. 2). In order to achieve an efficient and quick response, countries should ensure that any suspicion or confirmation of an outbreak should be immediately notified to the competent authority in order to ensure the timely implementation of necessary risk management measures. Such notification will also enable neighbouring or other affected countries to take precautionary measures to avoid the further spread of the disease. It is worth mentioning that disease control measures usually have a significant economic impact; therefore, appropriate compensation mechanisms should be in place to ensure cooperation by farmers. Lack of compensation could lead to non-compliance, which could mean not reporting the disease (Jurado et al., 2018).

3 Biosecurity measures

Pig production in the EU is highly heterogeneous regarding farm type (industrialized, outdoor or backyard), biosecurity standards and production levels (commercial production at different scales, backyard rearing solely for family consumption) (Bellini, 2021). Within Europe, large fattener farming operations rear 75% of all pigs but represent just 1.5% of all farms with pigs. Small-scale pig producers are concentrated in the 13 Member States of Eastern Europe that joined the EU in 2004 (Anon, 2014a) (Fig. 3). These countries were most affected by the last ASF epidemic in the EU in 2019. Smallholders generally have lower levels of farm biosecurity than commercial farms, and poor biosecurity is known to be a major factor in disease development and spread. Given the absence of effective vaccines, biosecurity is critical in preventing the spread of ASF (Blome et al., 2020). A high percentage of smallholders in the domestic pig sector is considered an important indicator of the potential spread of ASF due to both poor biosecurity measures and other features that are typical of this type of farming systems, such as swill feeding, illegal animal movements, collection of backyard pigs in markets and home slaughtering (Anon, 2019d). Backyard units that sell animals at local markets have played a significant role in the local spread of ASF and have contributed to it becoming endemic in some regions (Bellini, 2018).

The World Organization for Animal Health (OIE) conducted a recent survey on how biosecurity is applied across Europe to analyse current strengths and weaknesses and identify best practices (Bellini, 2018). The study revealed that the poultry and pig farming sectors demonstrate the most frequent use of

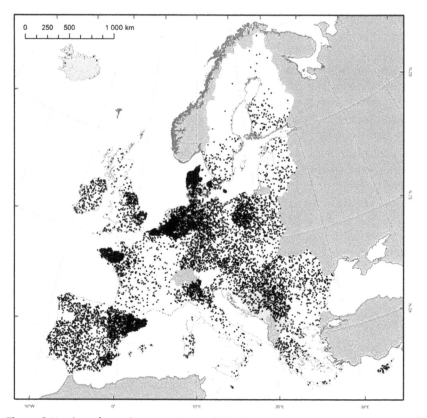

Figure 3 Number of sows by region (Anon, 2014a).

biosecurity practices. It also showed that biosecurity is most rigorously applied in commercial farms, even though all the holdings that have access to markets, including backyards, should be included in biosecurity planning.

The Working Document of DGSANTE 'Strategic Approach to the management of African swine fever for the EU SANTE/7113/2015-Rev 12' (Anon, 2015) has been developed in response to the current prevalence of ASF in Europe and seeks to establish harmonized measures to prevent and control the spread of ASF in the affected territories of the EU. Based on this document, pig farms are classified into three categories:

- non-commercial farms where pigs are kept only for fattening for consumption by the owner and neither pigs nor any of their products leave the holding;
- commercial farms which sell pigs, send pigs to a slaughterhouse or move pig products off the holding; and
- outdoor farms where pigs are kept temporarily or permanently outdoors.

This classification does not take into account the size of the farm or the type of establishment (breeding, fattening, etc.). Instead, it focuses on the risk of spreading the disease by moving pigs and the risk of being exposed to an external source of infection, such as the presence of infected wild pigs or soft ticks. Farms are categorized based on the risk of spreading ASF, and biosecurity measures are recommended accordingly.

The risk of exposure to ASF for the individual farm depends on the country, area and farm location in relation to the infectious status in the surroundings. The proximity and number of ASF cases in domestic pigs and in wild boar were found to be significant risk factors for the introduction of ASF in pig farms in Romania (Boklund et al., 2020). This implies that biosecurity measures should take into account virus persistence in the environment and routes of transmission, as well as the characteristics of farming systems.

Farm biosecurity plans aimed at minimizing the risk of spreading pathogens should include the following elements (Anon, 2010a):

- Segregation/separation: The creation and maintenance of barriers to limit potential opportunities for infected animals and contaminated materials to enter a farm; when properly applied, this step will prevent most contamination and infection;
- Cleaning: Thorough cleaning of vehicles and equipment entering or leaving a farm to remove visible dirt. This practice will also remove most of any pathogens present; and
- Disinfection: When properly applied, disinfection will inactivate any pathogen that is present on materials that have already been thoroughly cleaned.

Jurado et al. (2018) conducted a systematic review of the documentation published on the measures to prevent the spread of ASF in domestic pigs. The study classified preventive measures into four different groups: general preventive measures (measures in common for commercial, non-commercial and outdoor farms) and three groups of suggested measures for each of the identified types of farms (Jurado et al., 2018). A group of experts was subsequently invited to participate in an expert opinion session to develop a comprehensive list of biosecurity measures aimed at minimizing the risk of ASFV introduction in pig farms, taking into account the requirements described in the DGSANTE working document (Anon, 2015) (Tables 1 to 3). All the experts agreed that the following measures were relevant for ASF prevention on all three types of pig farms:

- strict enforcement of the ban on swill feeding;
- containment of pigs to avoid contact with pigs from other farms, feral pigs or wild boar; and
- good identification of individual animals and records of animal movements.

Table 1 ASF biosecurity guidance for commercial farms (Martínez et al., 2021)

Farm location	Far from suitable wild boar habitat, pig farms frequently used roads. Farm perimeter fenced (double fenced) and farm entrance closed.
Structural Biosecurity	• Loading and unloading areas placed at least 20 m away from animal facilities within the perimeter of the farm, built with materials easy to clean and disinfect and 1% negative inclination. Drivers should not come in contact with farm workers/animals. Establish clear cleaning and disinfection procedures for vehicles. • Quarantine rooms for new animals (30 days). Animals inspected by veterinarians (serologically and virologically testing when necessary). Isolation rooms for infectious/sick animals. • Nets against birds/insects. Pest control periodically conducted. The presence of domestic animals should be discouraged. • Handwashing facilities/changing rooms/showers. Footbaths at the entrance of every animal unit. Organic material should be removed from footwear prior to disinfecting. • Use of materials in structures or buildings that facilitate cleaning and disinfection (C&D) procedures.
Animal health status	Health status and ASF-free certificates have to be checked before introducing onto the farm: pigs, semen, ova and embryos. Buying pigs at local markets and/or illegally is a high-risk practice.
All-in-all-out system	C&D protocols after moving out pigs + fallow period. All organic material should be completely removed to maximize the efficacy of disinfection.
Good farming practices	• C&D protocols should be established and periodically performed on every farm facility, vehicle and tool. • Workers and farm owners should follow the described biosecurity measures with regard to handwashing, showering, changing of clothes and footwear. They should not have swine herds or pig pets at home. • No food should be brought onto animal facilities. • Sharing material between animals, especially needles, should be avoided. Any equipment like ultrasound apparatus that usually is owned by the veterinarian and moved from farm to farm should be cleaned and disinfected before entering the farm. • Straw and bedding materials should come from ASF-free areas where no slurry has been used as fertilizer, and there is no possibility of contamination with ASF-infected wild boar. • Manure and dead animals should be properly disposed. Containers should be located far from animal facilities. Vehicles should have access to these containers without entering the farm. • To avoid cross-contamination, proper management and storage of animal feed should be done. • The presence of pets should be discouraged in pig facilities or where feed is stored.
Farm visits	As a general rule, they should be discouraged, when necessary must be recorded. Veterinarians should follow strict biosecurity measures, and their visit should be recorded on the farm record book. Veterinary instruments and equipment should be properly cleaned and disinfected. Workers and owners should also avoid visiting other pig farms.
Awareness and training of farm staff	Should be raised and updated according to the level of risk.
Farm records	Births and deaths, animal census and movements, visits, pest control and C&D procedures shall be registered in the farm record book, as well as the management protocols for each production phase.

Table 2 ASF biosecurity guidance for non-commercial farms (Martínez et al., 2021)

1	Buying pigs from trusted and certified sources (ASF virus-free commercial holdings).
2	No swill feeding.
3	Ensure identification of holdings, pigs and their traceability.
4	Workers should not bring food onto the premises.
5	The access to pig's stable should be restricted only to people in charge of taking care of the animals.
6	Pig owners and people in charge of pig shall avoid visiting other farms.
7	Pigs shall be kept confined in stables: pigs should be kept in a way that ensures that there is no direct, or indirect, contact with other pigs outside the premises nor with wild boar.
8	Farm building should:
	a be built in such a way that no feral pigs or other animals (e.g. dogs) can enter the stable.
	b allow for disinfection facilities (or changing) for footwear at the entrance into the stable.
9	Effective disinfectants should be placed at the entrance of the stable.
10	No sows or boar used for natural reproduction shall be allowed on the holding. In case pigs intended for breeding are kept on the holding, biosecurity should follow the same indications foreseen for commercial farms.
11	People working in contact with pigs should wear clothes and footwear to be worn and used only when working in the stable and to be left in the stable after use.
12	People working in contact with pigs should wash hands with soap before entering and leaving the premises.
13	Proper disposal of dead animals or parts of dead animals should be done to avoid the spread of infected material and also to attract wild animals.
14	No wild boar or part of it shall be brought onto the premises.
15	No hunting activity should be carried out 48 h prior to being in contact with pigs.

If locally harvested grass and straw are used in areas at risk:
1. Ban of feeding fresh grass/grain to pigs unless treated to inactivate the ASF virus or stored for at least 30 days before feeding.
2. Ban on using straw for bedding of pigs unless treated to inactivate ASF virus or stored (out of reach of wild boar) for at least 90 days before use.

Enforcing these basic measures on all farms is key to limiting the incidence and spread of ASF.

It is important to note that, although it is a known transmission route for ASF, swill feeding is a common practice in traditional pig production systems involving free-range and backyard pigs (Anon, 2019d). Because it contains animal matter, swill can harbour ASFV, which can persist for months in pork meat, fat and skin and in different types of pork products (McKercher et al., 1987; Mebus et al., 1993, 1997; Anon, 2009; Gogin et al., 2013). Since swill may contain other viruses causing diseases such as foot-and-mouth disease and

Table 3 ASF biosecurity guidance for outdoor pig-production systems (Martínez et al., 2021)

Measures to reduce the risk of ASF within the farm	Measures to reduce the risk of ASF entrance to the farm
No swill/catering waste feeding	Farm location should be far away from areas with wild boar or other pigs and main roads.
Entry of new animals, same biosecurity measures as indoor commercial pigs (including quarantine).	Fences to avoid direct transmission from potential infected wild boar/feral/free-ranging suids. Double fencing in outdoor holdings if located close to wild boar suitable habitat or other solution that excludes the possibility of contact with free wildlife.
Safe and controlled disposal of dead animals in rendering plant + waste/residues management (including slurry) to avoid the spread of infected material or to attract wild animals.	Control of feed: No swill/catering waste feeding Control of water points to avoid accessibility by other animals. Higher surveillance at times of the year when resources are scarce in nature.
Avoid the use of items that increase the contact among pigs, or management practices that would induce fights.	C&D vehicles, equipment, weeds, etc.
Daily visit to check pig health status.	Treatment of any bedding material. Avoid usage of freshly mowed green mass (grass and grain) and straw.
Appropriate infrastructure to avoid tick breeding.	Workers' biosecurity practices: Avoid visiting other pig farms, hunting, etc.; clothing; hygiene (hand washing); food management.
Use of acaricides and insecticides.	Control of visits and other people presence
No other animal species in outdoor pig holding.	No sharing of feeding, equipment, etc.
	Avoid entry of vehicles to the place where animals are kept.
	Biosecurity requirements controlled in each visited holding.

classical swine fever, as well as ASF, swill feeding is banned in the EU. Article 15.1.22 of the OIE Terrestrial Code recommends heat treatment as a means to mitigate the risk of spreading ASFV throughout swill feeding.

Direct contact with infected animals or infectious material is the other area of risk highlighted by the experts. At the farm level, special attention shall be given to the management of animal transport, the C&D of vehicles and the loading/unloading area. Before introducing pigs onto the farm, it is necessary to verify the health status of the supplier, which should never be less than that of the purchasing farm. Suppliers should be few, trusted (with recognized certification for animal health) and ASF-free. Pigs must be transported in accordance with relevant registration and identification rules, and all pigs in the vehicle should be intended for a single holding; vehicles must be cleaned and disinfected before loading the pigs and after unloading (Bellini, 2015).

New pigs entering the farm should first be isolated and quarantined. The quarantine area should be physically separated from the rest of the farm and be thoroughly cleaned and disinfected before new pigs are introduced. The quarantine area should have dedicated staff and be managed following the all-in/all-out system. A minimum of 30 days of quarantine is recommended in the case of ASF. During the quarantine period, animals should be carefully checked to detect the early presence of infections in order to avoid the introduction of diseased animals into the herd (Bellini, 2015). As noted, clinical surveillance is the most effective tool for the early detection of ASF. However, given the clinical similarity with other diseases of pigs, clinical surveillance should be supplemented, as appropriate, by serological and virological surveillance (Anon, 2019b).

An analysis carried out by EFSA (Anon, 2019d) suggests that both smallholder and free-range pig farming often have a low level of compliance with legal requirements on the registration and identification of animals, reducing the effectiveness of tracking and managing the disease. The introduction and maintenance of these identification and tracking systems is a legal requirement for EU member states. All pigs should be identified, and all animal movements should be recorded. Births and deaths, entry and exit of animals (live and dead), transport details, visits, pest control and C&D procedures should all be properly recorded. Protocols for each production phase should be clearly defined, including movement flows within the farm.

The need for proper identification and tracking systems, as well as properly documented and implemented biosecurity measures, is now supported by online information and teaching tools, mobile phone data collection and epidemiological decision support systems to facilitate the collection and transformation of information, training of professionals and the management of veterinary activities. Ghent University of Veterinary Medicine (Belgium) has developed a scoring system (Biocheck.UGent) to evaluate biosecurity in poultry and pig farms. This tool covers all relevant components of a biosecurity system. The system assesses the relative importance of the different elements of biosecurity and produces a risk-based weighted score. The scoring system is supported by a questionnaire that can be completed online. The system returns an individual biosecurity score that the farmer can compare with national average values. The system also provides tailored advice on how biosecurity could be improved on the farm. A similar system has been developed by the Swedish pig industry (www.smittsäkra.se). The website contains advice on disease prevention, detailed information on infectious diseases and how they may be spread, as well as self-testing to assess a farm's level of disease protection. The website is run by farmer veterinary organizations in cooperation with the National Veterinary Institute and is financed by the Board of Agriculture (Bellini, 2018). Research indicates that many personnel involved in pig farming have a poor understanding of biosecurity, the exception being larger, intensive commercial operations (mainly pigs, poultry and ruminant feedlots); this situation could be improved through appropriate training and awareness campaigns (Windsor, 2017).

4 Cleaning and disinfection measures

Along with biosecurity measures, pathogen inactivation through proper C&D procedures is an essential step both to prevent ASF spreading and to facilitate recovery after an outbreak (Ford, 1995). The EU Commission Delegated Regulation 2020/687/ establishes that, in the case of ASF, the destruction of carcasses should be followed by thorough C&D of all premises, vehicles and equipment. These operations should be conducted under the supervision of the relevant veterinary authority and, where official approval is required, be certified by the supervising official veterinarian. The OIE requires a complete C&D cycle in order to regain ASF-free status after an outbreak (Anon, 2019b).

A complete C&D protocol consists of seven steps:

- dry cleaning;
- wet cleaning;
- rinsing;
- drying;
- disinfection;
- drying; and
- testing the efficiency of the procedure.

When properly applied, a complete C&D protocol removes over 90% of microorganisms and improves disinfection efficacy (Anon, 2014b) (Table 4). Complete procedures for C&D have been reviewed by De Lorenzi et al. (2020).

Table 4 Detergent/disinfectants not to be used in combination

Disinfectant	Detergent	Cause
Quaternary ammonium compounds (QACs)	Alkalis	Alkaline detergents may react chemically with QACs and destroy their antimicrobial properties
Hypochlorite	Acids detergents	If these compounds are mixed, the resultant reaction releases toxic chlorine gas
Phenols	Soaps based on tallow, tall oil or oleic acids	These detergents are able to markedly decrease the activity of phenol compounds
Chlorhexidine	Alkalis	Alkaline detergents may interfere with disinfectant action of chlorhexidine

Source: Holah (1992, 1995); Turner and Burton (1997); Turner et al. (1999).

Table 5 Disinfectants effective against ASFV

Active ingredient(s)	Contact time	Application(s)
Sodium chloride Potassium peroxymonosulfate	10 min	In/on animal feeding equipment, livestock barns/pens/stalls/stables, livestock equipment, hog farrowing pen premises, hog barns/houses/pens, animal quarters, animal feeding and watering equipment, animal transportation vehicles, agricultural premises/equipment, human footwear
Sodium dichloro-s-triazinetrione	30 min	In/on animal living quarters, farm premises, shoe baths
Sodium dichloro-s-triazinetrione	30 min	In/on animal living quarters, farm premises, shoe baths
Sodium dichloro-s-triazinetrione	30 min	Animal quarters, animal feeding/watering, animal equipment, transportation vehicles
Sodium dichloro-s-triazinetrione	30 min	In/on livestock premises, animal feeding/watering equipment, animal equipment, animal transportation vehicles, farm premises, shoe baths
Sodium hypochlorite	15 min nonporous 30 min porous	Indoor or outdoor use sites, such as agricultural, transportation, quarantine, and laboratory equipment and facilities; footwear/personal protective equipment
Citric acid	15 min nonporous 30 min porous	Indoor or outdoor use sites such as agricultural and non-agricultural equipment and facilities; laboratory equipment and facilities; footwear/personal protective equipment, personnel decontamination

Source: Anon (2011 modified); Geering et al. (2001).

In areas affected or at risk of introducing ASF, the disinfectants chosen must be effective against the ASFV and approved by the official veterinarian. The choice of disinfectants and procedures for disinfection must take into consideration the nature of the premises, vehicles, objects to be treated (Table 5) and the conditions for their use strictly adhered to. A list of effective chemicals and disinfectants has been compiled, though the use of some of them is limited by their toxicity (De Lorenzi et al., 2020). There is no single ideal disinfectant against ASFV, but every country has approved and/or authorized a list of biocides effective against ASFV; only authorized biocides should be used and applied according to the producer's instructions (Juszkiewicz, 2019). The OIE website lists effective disinfectants against ASFV (Anon, 2019c).

There is a significant body of knowledge of and experience in the use of disinfectants against enveloped viruses (Anon, 2004, 2019c; Gallina and Scagliarini, 2010; Krug et al., 2011, 2012; Shirai et al., 2000; Stone and

Hess, 1973; Turner and Burton, 1997). This research has shown that chemical compounds effective in the inactivation of ASFV are:

- formaldehyde 1%;
- sodium hypochlorite (0.03–0.0075%);
- caustic soda solution 2%;
- glutaraldehyde, formic acid;
- sodium or calcium hydroxide 1% (effective at virus inactivation in slurry at 4°C);
- phenols – lysol, lysephoform, and creolin;
- chemical compounds based on lipid solvents; and
- multi-constituent compounds – sodium chloride, potassium peroxymonosulfate, lysoformin, desiform, octyldodeceth-20 (OD-20) surfactants, active substances, organic acids, glycosal, etc.

Disinfectants recommended in the United States by the Environmental Protection Agency (EPA) and the conditions for their use are shown in Table 5. Currently, there are commercial disinfectants based on phenolic and iodine compounds which are effective against the virus and can inactivate the ASFV at pH < 4 and > 11 (Gallardo et al., 2015; Geering et al., 2001).

The choice of the disinfectant must take into consideration aspects such as the nature of the premises, vehicles and objects to be treated (Table 6), types of surface, the spectrum of activity, the efficacy and practicability under farm conditions (e.g. ease of handling, risk of corrosion of equipment, temperature stability); safety towards humans, animals and the environment; costs; storage requirements; etc. (Anon, 2010). The preparation of disinfectant solutions must be performed by qualified operators following the manufacturer's instructions (concentration, contact time, pH, temperature) (Anon, 2010). All disinfectants, whether they are sprays, foams, aerosols or fumigants, work best at a temperature above 18.3°C, whereas temperatures for chlorine

Table 6 List of disinfectants incompatible with metal surfaces

Chemical disinfectant	Effect on metal surfaces
Sodium hydroxide	Corrosive to aluminium and derived alloys and galvanized metal
Sodium carbonate	Corrosive to aluminium and derived alloys
Acids	Highly corrosive to metals
Glutaraldehyde, Virkon® S	Mildly corrosive to metals
Iodophors, hypochlorites, formaldehyde	Corrosive to some metals
Phenolics	Relatively non-corrosive

Source: Anon (2014b).

and iodine-based disinfectants should not exceed 43.3°C (Anon, 2018). It is important to keep solutions clean and freshly made. Mixing disinfectants is inadvisable since each disinfectant has optimal pH to maximize effectiveness that can be altered by mixing different substances (Anon, 2004). After application, disinfectants should be left on the surface to act for an appropriate contact time according to the manufacturer's instructions on the label. In the case of ASF, disinfectants must remain on the surfaces for at least 24 h (Anon, 2020). Contact time can be increased by applying the disinfectant as a foam or gel, in which case contact times of 10–15 minutes (foam) and >15 minutes (gel) are possible.

5 Conclusion

The worldwide spread of ASFV and its dramatic impact on pork production resulted in several studies to identify the risk factors involved in the spread of ASFV to domestic pigs (Bellini, 2021). This information is of crucial importance in planning targeted disease control measures aimed at early detection and preventing the spread of ASF. Many ASF outbreak investigations have reported biosecurity shortcomings as a critical element for virus introduction and spread. The structure of the European swine industry makes it necessary to put in place differentiated biosecurity measures to meet the different risk levels for the introduction and spread of ASF among diversified farming systems. Biosecurity is often targeted at commercial holdings, whereas all holdings that have access to markets should be included in a biosecurity programme. Backyard units that sell animals at the local or regional level can have a role in the spread of disease.

The final responsibility in the control of ASF belongs to veterinary authorities. However, effective prevention and control of ASFV infection require input from a wide range of stakeholders involved in pork production. National animal health authorities have a key role in developing and implementing appropriate policy instruments, biosecurity regulations, surveillance strategies and outbreak response policies. The latter are typically based on the current understanding of the epidemiology of the disease. However, they often do not explicitly take socioeconomic or cultural factors into account. This is a problem since livestock populations are managed by individuals who need to understand and support official initiatives (Loi et al., 2019). This is particularly important given evidence of human-mediated transportation of ASFV in some affected countries and as an important pathway contributing to the spread of the disease, for example, in Belgium, the Czech Republic and western Poland, as reported by EFSA (Anon, 2019d). Well-structured communication and awareness campaigns are needed for tourists, hunters, farmers, etc., to limit the risk of spread via movements of people, as well as to increase the probability of early detection.

6   Where to look for further information

Anyone wishing to gain knowledge on different aspects of ASF can consult the EFSA website (https://www.efsa.europa.eu/en/topics/topic/african-swine-fever), which can be considered a complete source of up-to-date information on the disease.

Updated information on the evolution of the epidemiological situation and on the disease control measures can be found in the World Organization for Animal Health, the OIE website (https://www.oie.int). The site reports also the contact persons of the OIE reference laboratories designated to provide scientific and technical assistance and expert advice on topics linked to diagnosis and control of the disease.

Furthermore, there are several current EU research projects on ASF, including:

- DEFEND (Addressing the dual emerging threats of African swine fever and lumpy skin disease in Europe). The project aims at identifying the drivers for emergence of ASF in Europe and to develop tools for its management.
- VACDIVA: which aims to provide an effective and safe vaccine(s) for wild boar and domestic pigs and to develop DIVA test to allow an accurate monitoring of the effectiveness of the vaccine.
- SWINOSTICS (Swine diseases field diagnostics toolbox) with the objective to develop a novel field diagnostic device, based on advanced, proven, bio-sensing technologies to tackle viruses.

Whereas the control measures to adopt to fight ASF in the European Union can be found at EU website (https://ec.europa.eu/food/animals/animal-diseases/diseases-and-control-measures/african-swine-fever_en).

7   References

Anon. (2004). Manuale operativo Peste Suina Classica e Peste Suina Africana. Available at: http://www.izsum.it/files%5CDownload%5C48%5C-1%5CManuale%20Operativo%20PSC-PSA.pdf. (accessed 06 April 2021).

Anon. (2009). Scientific report submitted to EFSA prepared by Sánchez-Vizcaíno, J.M., Martínez-López, B., Martínez-Avilés, M., Martins, C., Boinas, F., Vial, L., Michaud, V., Jori, F., Etter, E., Albina, E. and Roger, F. on African Swine Fever. 1–141.

Anon. (2010a). Food and agriculture organization of the United Nations/world organization for animal health/world bank. Good practices for biosecurity in the pig sector – issues and options in developing and transition countries. *FAO Animal Production and Health Paper*. Food and Agriculture Organization, Rome, 69, 169.

Anon. (2010b). Good practices for biosecurity in the pig sector – issues and options in developing and transition countries. *FAO Animal Production and Health Paper*. Food and Agriculture Organization of the United Nations (FAO)/World Organisation for Animal Health (OIE)/World Bank, 169. Available at: http://www.fao.org/3/i1435e/i1435e00.htm. (accessed 11 April 2021).

Anon. (2014a). EUROSTAT – statistics in focus 15/2014. Pig farming in the European Union: considerable variations from one Member State to another. ISSN: 2314-9647.

Anon. (2014b). FAD PReP/NAHEMS - NAHEMS guidelines: cleaning and disinfection. The Foreign Animal Disease Preparedness and Response Plan/National Animal Health Emergency Management System, Ames, IA, Riverdale, MD. Available at: https://www.aphis.usda.gov/animal_health/emergency_management/downloads/nahems_guidelines/cleaning_disfection.pdf. (accessed 02 April 2021).

Anon. (2014c). OIE guidelines for animal disease control. Available at: https://www.oie.int/fileadmin/Home/eng/Our_scientific_expertise/docs/pdf/A_Guidelines_for_Animal_Disease_Control_final.pdf. (accessed 30 March 2021).

Anon. (2015). Strategic approach to the management of African swine fever for the EU. SANTE/7113/2015-Rev 12. European Union Directorate General for Health and Food Safety, Brussels, Belgium. Available at: http://ec.europa.eu/food/sites/food/files/animals/docs/ad_control-measures_asf_wrk-doc-sante-2015-7113.pdf. (accessed 02 April 2021).

Anon. (2016). Regulation (EU) 2016/429 of the European Parliament and of the Council of 9 March 2016 on transmissible animal diseases and amending and repealing certain acts in the area of animal health ('Animal Health Law'). *Official Journal of the European Union* 84, 31.3.2016, 1–208.

Anon. (2018). FAD PReP/USDA. *Standard Operating Procedures: 15. Cleaning and Disinfection*. The Foreign Animal Disease Preparedness & Response Plan/ United States Department of Agriculture, Riverdale, MD. Available at: https://www.aphis.usda.gov/animal_health/emergency_management/downloads/sop/sop_cd.pdf. (accessed 31 March 2021).

Anon. (2019a). European Commission (EC), EU agricultural outlook for market and income 2019–2030. Directorate General Agriculture and Rural Development (DGAGRI), Brussels. Available at: https://ec.europa.eu/info/news/eu-agricultural-outlook-2019-2030-african-swine-fever-continues-impact-global-meatmarket-2019-dec-10_en. (accessed 02 April 2021).

Anon. (2019b). Terrestrial animal health code. Office International des Epizooties, Paris. Available at: https://www.oie.int/standard-setting/terrestrial-code/access-online/. (accessed 06 April 2021)

Anon. (2019c). African swine fever: aetiology epidemiology diagnosis prevention and control references, technical disease cards. Office International des epizooties, Paris. Available at: https://www.oie.int/fileadmin/Home/eng/Our_scientific_expertise/docs/pdf/AFRICAN%20SWINE%20FEVER.pdf. (accessed 06 April 2021).

Anon, Nielsen, S. S., Alvarez, J., Bicout, D., Calistri, P., Depner, K., Drewe, J. A., Garin-Bastuji, B., Gonzales Rojas, J. L., Michel, V., Miranda, M. A., Roberts, H., Sihvonen, L., Spoolder, H., Ståhl, K., Viltrop, A., Winckler, C., Boklund, A., Bøtner, A., Gonzales Rojas, J. L., More, S. J., Thulke, H. H., Antoniou, S. E., Cortinas Abrahantes, J., Dhollander, S., Gogin, A., Papanikolaou, A., Gonzalez Villeta, L. C. and Gortázar Schmidt, C. (2019d). Risk assessment of African swine fever in the south-eastern countries of Europe. *EFSA Journal. European Food Safety Authority* 17(11), e05861. https://doi.org/10.2903/j.efsa.2019.5861.

Anon. (2019e). Commission Delegated Regulation (EU) 2020/689 of 17 December 2019 supplementing Regulation (EU) 2016/429 of the European Parliament and of the Council as regards rules for surveillance, eradication programmes, and disease-free status for certain listed and emerging diseases. *Official Journal of the European Union* 174, 3.6.2020, 211–340.

Anon. (2020). Commission delegated. Regulation (EU) 2020/687 of 17 December 2019 supplementing Regulation (EU) 2016/429 of the European Parliament and the Council, as regards rules for the prevention and control of certain listed diseases. *Official Journal of the European Union* 174, 3.6.2020, 64–139 (accessed 27 September 2021).

Bellini, S. (2015). Standing Group of Experts on African swine fever in the Baltic and Eastern Europe region - SGE ASF2 focussed on biosecurity in pig production systems (including backyard) as a ASF control measure. GF-TADs, Tallinn, Estonia. Available at: https://web.oie.int/rr-europe/eng/regprog/en_gf_tads%20-%20standing%20group%20asf.htm#SGE2. (accessed 11–12 February 2015).

Bellini, S. (2018). Application of biosecurity in different production systems at individual, country and regional levels. O.I.E. World Organisation for Animal Health.

Bellini, S. (2021). The pig sector in the European Union. In: Iacolina, L., Penrith, M-L., Bellini, S., Chenais, E., Jori, F., Montoya, M., Ståhl, K. and Gavier-Widén, D. (Eds), *Understanding and Combatting African Swine Fever*. Wageningen Academic Publisher. https://doi.org/10.3920/978-90-8686-910-7_7.

Bellini, S., Rutili, D. and Guberti, V. (2016). Preventive measures aimed at minimizing the risk of African swine fever virus spread in pig farming systems. *Acta Veterinaria Scandinavica* 58(1), 82. https://doi.org/10.1186/s13028-016-0264-x.

Bellini, S., Scaburri, A., Tironi, M. and Calò, S. (2020). Analysis of risk factors for African swine fever in Lombardy to Identify Pig Holdings and areas most at risk of introduction in order to plan preventive measures. *Pathogens* 9(12), 1077.

Blome, S., Franzke, K. and Beer, M. (2020). African swine fever – a review of current knowledge. *Virus Research* 287, 198099.

Boklund, A., Dhollander, S., Chesnoiu Vasile, T., Abrahantes, J. C., Bøtner, A., Gogin, A., Gonzalez Villeta, L. C., Gortázar, C., More, S. J., Papanikolaou, A., Roberts, H., Stegeman, A., Ståhl, K., Thulke, H. H., Viltrop, A., Van der Stede, Y. and Mortensen, S. (2020). Risk factors for African swine fever incursion in Romanian domestic farms during 2019. *Scientific Reports* 10(1), 10215. https://doi.org/10.1038/s41598-020-66381-3.

Costard, S., Wieland, B., de Glanville, W., Jori, F., Rowlands, R., Vosloo, W., Roger, F., Pfeiffer, D. U. and Dixon, L. K. (2009). African swine fever: how can global spread be prevented? *Philosophical Transactions of the Royal Society of London. Series B, Biological Sciences*. Biological Sciences 364(1530), 2683–2696.

De Lorenzi, G., Borella, L., Alborali, G. L., Prodanov-Radulović, J., Štukelj, M. and Bellini, S. (2020). African swine fever: a review of cleaning and disinfection procedures in commercial pig holdings. *Research in Veterinary Science* 132, 262–267.

Dixon, L. K., Stahl, K., Jori, F., Vial, L. and Pfeiffer, D. U. (2020). African swine fever epidemiology and control. *Annual Review of Animal Biosciences* 8, 221–246.

Faverjon, C., Meyer, A., Howden, K., Long, K., Peters, L. and Cameron, A. (2020). Risk-based early detection system of African Swine Fever using mortality thresholds. *Transboundary and Emerging Diseases* 00, 1–11.

Ford, W. B. (1995). Disinfection procedures for personnel and vehicles entering and leaving contaminated premises. *Revue Scientifique et Technique* 14(2), 393–401.

Gallardo, M. C., Reoyo, A. T., Fernández-Pinero, J., Iglesias, I., Muñoz, M. J. and Arias, M. L. (2015). African swine fever: a global view of the current challenge. *Porcine Health Management* 23, 1–21.

Gallina, L. and Scagliarini, A. (2010). Virucidal efficacy of common disinfectants against orf virus. *Veterinary Record* 166(23), 725–726.

Geering, W. A., Penrith, M. L. and Nyakahuma, D. (2001). *Manual on Procedures for Disease Eradication by Stamping Out*. FAO Health Manual 12, Rome.

Gogin, A., Gerasimov, V., Malogolovkin, A. and Kolbasov, D. (2013). African swine fever in the North Caucasus region and the Russian Federation in years 2007–2012. *Virus Research* 173(1), 198–203.

Guberti, V., Khomenko, S., Masiulis, M. and Kerba, S. (2016). *GF-TADs Handbook on ASF in Wild Boar and Biosecurity during Hunting* (25.09.2018 version). Available at: https://web.oie.int//rr-europe/eng/eng/regprog/docs/docs/gf-tads%20handbook_asf_wildboar%20version%202018-09-25.pdf. (accessed 15 June 2021).

Guinat, C., Gogin, A., Blome, S., Keil, G., Pollin, R., Pfeiffer, D. U. and Dixon, L. (2016). Transmission routes of African swine fever virus to domestic pigs: current knowledge and future research directions. *Veterinary Record* 178(11), 262–267, https://doi.org/10.1136/vr.103593.

Holah, J. T. (1992). Cleaning and disinfection. In: Dennis, C. and Stringer, M. (Eds), *Chilled Foods: a Comprehensive Guide*. Ellis Horwood, London. pp. 319–341.

Holah, J. T. (1995). Disinfection of food production areas. *Revue Scientifique et Technique* 14(2), 343–363.

Loi, F., Laddomada, A., Coccollone, A., Marrocu, E., Piseddu, T., Masala, G., Bandino, E., Cappai, S. and Rolesu, S. (2019). Socio-economic factors as indicators for various animal diseases in Sardinia. *PLoS ONE* 14(6), e0217367, https://doi.org/10.1371/journal.pone.0217367.

Jurado, C., Martínez-Avilés, M., de la Torre, A., Štukelj, M., Cardoso de Carvalho Ferreira, H. C., Cerioli, M., Sánchez-Vizcaíno, J. M. and Bellini, S. (2018). Relevant measures to prevent the spread of African swine fever in the European Union domestic pig sector. *Frontiers in Veterinary Science* 5, 77. https://doi.org/10.3389/fvets.2018.00077.

Juszkiewicz, M., Walczak, M. and Woźniakowski, G. (2019). Characteristics of selected active substances used in disinfectants and their virucidal activity against ASFV. *Journal of Veterinary Research* 63(1), 17–25.

Krug, P. W., Lee, L. J., Eslami, A. C., Larson, C. R. and Rodriguez, L. (2011). Chemical disinfection of high-consequence transboundary animal disease viruses on nonporous surfaces. *Biologicals* 39(4), 231–235.

Krug, P. W., Larson, C. R., Eslami, A. C. and Rodriguez, L. L. (2012). Disinfection of foot-and-mouth disease and African swine fever viruses with citric acid and sodium hypochlorite on birch wood carriers. *Veterinary Microbiology* 156(1–2), 96–101.

Martínez, M., de la Torre, A., Sánchez-Vizcaíno, J. M. and Bellini, S. (2021). Biosecurity measures against African swine fever in domestic pigs. In: Iacolina, L., Penrith, M-L., Bellini, S., Chenais, E., Jori, F., Montoya, M., Ståhl, K. and Gavier-Widén, D. (Eds), *Understanding and Combatting African Swine Fever*. Wageningen Academic Publisher. https://doi.org/10.3920/978-90-8686-910-7_7.

McKercher, P. D., Yedloutshnig, R. J., Callis, J. J., Murphy, R., Panina, G. F., Civardi, A., Bugnetti, M., Foni, E., Laddomada, A., Scarano, C. and Scatozza, F. (1987). Survival of viruses (Parma ham). *Canadian Institute of Food Science and Technology Journal* 20(4), 267–272.

Mebus, C. A., House, C., Ruiz Gonzalvo, F. R., Pined, J. M., Tapiador, J., Pire, J. J., Bergada, J., Yedloutschnig, R. J., Sahu, S., Becerra, V. and Sánchez-Vizcaíno, J. M. (1993). Survival of foot-and-mouth disease, African swine fever, and hog cholera viruses in Spanish Serrano cured hams and Iberian cured hams, shoulders and loins. *Food Microbiology* 10(2), 133–143. https://doi.org/10.1006/fmic.1993.1014.

Mebus, C., Arias, M., Pineda, J., Tapiador, J., House, J. and Sánchez-Vizcaíno, J. M. (1997). Survival of several porcine viruses in Spanish dry-cured meat products. *Food Chemistry* 59, 555–559. Available at: https://doi. https://doi.org/10.1016/S0308 -8146(97)00006-X.

Mur, L., Atzeni, M., Martinez-Lopez, B., Feliziani, F., Rolesu, S. and Sanchez-Vizcaino, J. M. (2016). Thirty-five year presence of African swine fever in Sardinia: history, evolution and risk factors for disease maintenance. *Transboundary and Emerging Diseases* 63(2), e165–e177. https://doi.org/10.1111/tbed.12264.

Pfeiffer, D. U., Ho, H. P. J., Bremang, A., Kim, Y. and OIE team (2021). *Compartmentalisation Guidelines – African Swine Fever*. World Organization for Animals Health (OIE), Paris. France. 148 pp.

Plavšic, B., Rozstalnyy, A., Park, J. Y., Guberti, V., Depner, K. R. and Torres, G. (2019). Strategic challenges to global control of African swine fever. 87th General Sessions on the World Assembly of the Delegates of the OIE, Paris, 26–31 May 2019.

Shirai, J., Kanno, T., Tsuchiya, Y., Mitsubayashi, S. and Seki, R. (2000). Effects of chlorine, iodine, and quaternary ammonium compound disinfectants on several exotic disease viruses. *Journal of Veterinary Medical Science* 62(1), 85–92.

Stone, S. S. and Hess, W. R. (1973). Effects of some disinfectants on African swine fever virus. *Applied Microbiology* 25(1), 115–122.

Turner, C. and Burton, C. H. (1997). The inactivation of viruses in pig slurries: a review. *Bioresource Technology* 61(1), 9–20.

Turner, C., Williams, S. M. and Wilkinson, P. J. (1999). Recovery and assay of African swine fever and Swine vesicular disease viruses from pig slurry. *Journal of Applied Microbiology* 87(3), 447–453.

Wilson, A. J., Ribeiro, R. and Boinas, F. (2013). Use of a Bayesian network model to identify factors associated with the presence of the tick *Ornithodorus erraticus* on pig farms in southern Portugal. *Preventive Veterinary Medicine* 110(1), 45–53.

Windsor, P. A. (2017). How to implement farm biosecurity: the role of government and private sector. OIE Asia Regional Commission. Available at: http://www.oie.int/fileadmin/Home/eng/Publications_%26_Documentation/docs/pdf/TT/2017_ASI1 _Windsor.pdf. (accessed 11 April 2021).

World Organization for Animal Health (2021). World animal health information database (OIE-WAHID). Available at: https://wahis.oie.int/#/home. (accessed 07 April 2021).

Chapter 4

Advances in finding a vaccine for African swine fever

Douglas P. Gladue and Manuel V. Borca, Plum Island Animal Disease Center and Center of Excellence for African Swine Fever Genomics, USA

1 African swine fever virus proteins induce protective immunity and the development of subunit vaccines

2 Live-attenuated vaccines – naturally attenuated and cell culture passed field isolates

3 Live-attenuated vaccines – Initial identification of genetic determinants of virulence in African swine fever virus

4 Case study: development of an African swine fever virus vaccine for the new emerging strain African swine fever virus Georgia

5 Other advances in African swine fever virus discovery of determinants of virulence

6 Conclusion

7 Future trends in research

8 Where to look for further information

9 References

1 African swine fever virus proteins induce protective immunity and the development of subunit vaccines

African swine fever virus (ASFV) is a structurally complex virus that encodes for over 180 different viral proteins, some of them presenting amino acid sequence variability among different isolates. Partially due to a large number of viral proteins, the protective immunity induced by individuals or groups of ASFV proteins is unknown. This knowledge is typically required to rationally design a subunit vaccine. Several studies have been done trying to identify these protective antigens with, in general, very modest results.

Several recent attempts to use well-known ASFV antigenic proteins have been conducted in multiple research groups. However, these attempts are typically limited in the number of proteins that have been tested. For example, a combined DNA/protein vaccination strategy was attempted using recombinant proteins from the ASFV Armenia 2007 isolate expressed in

http://dx.doi.org/10.19103/9781786768629.04
Published by Burleigh Dodds Science Publishing Limited, 2024.

baculovirus (full-length *CP530R* gene for proteins p15 and p35, *E183L* for p54, and *B646L* for p17 from the *Ba71ASFV* isolate) along with DNA encoding for viral proteins p32(CP204L), CD2 (EP402R), p72 (B646L), and p 17 (D117L). This study, using seven ASFV proteins, did not offer any protection when vaccinated animals were challenged with the current pandemic strain [1]. A similar vaccine approach using BacMam, a baculovirus-based vaccine using virus proteins p54, p30, and an extracellular domain of the viral hemagglutinin, CD2, resulted in specific T-cell responses that were quickly lost, only offering partial protection against a low virulence strain of ASFV [2]. In another study, cocktails using replication-deficient adenoviruses expressing either ASFV antigens p32, p54, pp62 (CP530R), and p72, or A151R, B119L, B602L, CD2ΔPRR, B438L, K205R-A104R, pp62, p72, and pp220(CP2745L), induced an antigen-specific antibody, INF-γ, and cytotoxic T-lymphocyte responses [3, 4] but were unable to protect pigs from challenge with the virulent Georgia 2007 isolate. In another study, cocktails using adenovirus-delivered ASFV proteins EP153R, p10(K78R), p15, CP80R, I329L, H108R, K196R, CP312R, F334L, NP419L, NP868R, B66L, H339R, R298L, EP153R, p10, p15, CP80R, I329L, H108R, K196R, CP312R, F334L, NP419L, NP868R, B66L, H339R, and R298L were challenged with the Arm07 isolate, a derivative of the current pandemic strain Georgia07. In this study, none of the animals survived [5]; however, there was no measurable antibody response at the time of challenge, as the antibody response to this cocktail of ASFV proteins was short-lived.

A systematic approach to identify antigenic proteins in ASFV was conducted by using 40 ASFV proteins, using a DNA-prime immunization followed by a Vaccinia virus delivery system, to determine the immunogenicity of these proteins. Although this work reported a significant knowledge about the immunogenicity of these proteins, this strategy was not successful as a vaccine strategy and failed to protect animals after challenge [6]. Using this information, several viral-vectored pools were tested and one pool consisting of virus proteins B602L, B646L, CP204L, E183L, E119L, EP153R, F317L, and MGF505-5R resulted in protection from the virulent strain of ASFV (OUR T88/1), when these proteins were administered using an approach including adenovirus delivery to prime the animals and a Modified Vaccina Ankara (MVA) delivery system to boost the animals [7]. This promising result suggests that it is possible to make a subunit or vectored vaccine for ASFV. However, to date, there have been no reports of this success with a highly virulent strain of ASFV such as the current pandemic strain ASFV-Georgia 2007 or its derivative viruses.

Other studies have been done to evaluate the antigenicity of different ASFV viral proteins (A151R, B119L, B602L, EP402RΔPRR, B438L, K205R, and A104R) using the adenovirus delivery system and found to elicit a robust immune response, but challenge studies were not performed [8].

Published by Burleigh Dodds Science Publishing Limited, 2024.

In summary, the ASFV antigens responsible to induce protection are still largely unknown, and work is being continued in this area by several groups studying ASFV. As mentioned, although there have been some promising results for subunit vaccines against low virulence strains of ASFV, additional work has to be performed in this area to discover the ASFV proteins involved in protective immunity in both animals surviving outbreaks or in animals with immunity induced by experimental live-attenuated vaccines.

2   Live-attenuated vaccines – naturally attenuated and cell culture passed field isolates

Animals surviving the infection with a particular strain of ASFV usually became protected from the disease caused by the same strain or an antigenically related one [9-11]. This discovery led to the identification of attenuated strains as a potential vaccine. Attenuation of ASFV has been also achieved by cell culture passage, a methodology that has been shown to be effective for many different viruses. Animals that were infected with these attenuated field strains were shown to be resistant to the infection with homologous virulent strains of ASF. In the laboratory, ASFV is typically grown in primary swine macrophage cultures, as there is no established cell line that supports the virus growth. Adapting virulent field isolates to stable cell lines has typically resulted in a reduction of virulence in domestic pigs [12], with increasing attenuation occurring with additional passes in cell cultures. In most cases, the adaptation of ASFV to cell culture has resulted in large genomic deletions, although the exact mechanism or specific loss of genes that is required for this cell culture adaptation is still unknown. For the pandemic strain ASFV Georgia, the concept of adapting ASFV Georgia to grow in the Vero cell line was tested and was shown similarly to other cell-adapted strains of ASFV that large genomic deletions were present after adaption to cell cultures. In the adaptation of ASFV Georgia to Vero cells, deletions were observed at both ends of the genome, and additional passes in Vero cell cultures further decreased the virulence of this virus; however, the adapted virus did not offer any protection against virulent challenge [13]. The efforts to adapt ASFV to cell cultures have recently been reviewed in detail by Sereda et al. [14]. However, an adaption of ASFV to cell cultures to attenuate field isolates has not been generally accepted as a successful method to develop a safe and effective vaccine for ASFV.

The allure of these cell cultures passed attenuated field strains to control ASF was used in a massive vaccination campaign during a disease outbreak in Portugal in the 1960s. However, it was later discovered that the administration of this attenuated strain had negative consequences since a high proportion of the vaccinated animals developed chronic lesions and crippling arthritis. This residual virulence has hindered the use of all the naturally attenuated or cell culture attenuated strains that have been tested experimentally as vaccine candidates. However, these naturally attenuated field isolates have served a purpose and have been regularly used as experimental vaccines to gain information for ASF vaccinology, as they are able to induce protection against homologous challenges. All of these naturally attenuated field strains had some level of residual virulence. The residual virulence in the Portuguese attenuated field isolates OUR T88/3 [15-17] and NH/P68 [12] have been well characterized as well as the recently discovered Georgia derivative attenuated field strain isolated in Latvia (strain Lv17/WB/Rie1) [18]. These three strains are good examples of these types of attenuated ASF viruses. Some attempts have been made to decrease the residual of these strains by making genomic changes and deleting virus genes associated with virulence. The resulting recombinant viruses either retained their virulence [19, 20] or lost their protective efficacy [20]. Therefore, so far, the potential use of naturally attenuated isolates as vaccine candidates has been unsuccessful for further development toward a safe and effective ASF vaccine.

3   Live-attenuated vaccines – Initial identification of genetic determinants of virulence in African swine fever virus

Generation of recombinant ASFV viruses with deletions in genes that are determinants of virulence constitutes a rational approach to developing attenuated virus strains from field isolates. This approach began in the 1990s when the first determinants of virulence were discovered for ASFV. The methodology for the deletion of genes in an ASFV field isolate was reported for the first time describing the deletion of the *NL-S* gene of ASFV encoded by the open reading frame DP71L. This gene was discovered as having partial homology to the neurovirulence-associated gene of herpes simplex virus *ICP34*. Deletion of this gene from the ASFV E70 isolate resulted in the ASFV-E70-ΔNL-S strain, which was the first recombinant ASFV strain that had drastically reduced virus virulence in domestic pigs [21]. In this study, *ASFV-E70*-DNL-S was intramuscularly (IM) inoculated with 10^2 or 10^3 TCID$_{50}$, a tissue culture infectious dose. All animals remained clinically normal having a long viremia (2–3 weeks) and, importantly, all animals survived a 30-day post-vaccination (dpv) challenge with 10^2 TCID$_{50}$ of the parental E70 isolate without presenting any clinical signs associated with ASF. For the first time, the deletion of a genetic determinant of virulence in ASFV resulted in reduced virulence, and the resulting recombinant virus was able to protect against the homologous challenge of the parental strain. This methodology is the basis for all recombinant work in ASFV and showed the possibility of rationally developing a live-attenuated vaccine for ASFV based on specific deletions of the ASFV genome.

The *UK* (*DP96R*) gene [22] is located immediately upstream of the NL-S in the E70 isolate and, although highly conserved across many ASFV isolates, its function still remains largely unknown. However, this gene was deleted from the ASFV E70 isolate without causing adverse effects on growth in primary swine macrophages. Importantly, all animals inoculated with 10^2 TCID$_{50}$ IM of an ASFV E70 lacking the *UK* gene (ΔUK) survived the infection but presented a transient period of fever and lethargy right after inoculation. Animals infected with the ΔUK virus showed long viremias (4-6 weeks) and survived the IM challenge (by day 42 pv) with 10^4 TCID$_{50}$ of the homologous virulent E70 isolate, remaining clinically normal during the observation period.

The Malawi Lil-20/1 isolate was first used as a backbone for deleting the thymidine kinase (TK) gene from ASFV. The TK gene deletion was observed to impede replication of ASFV in swine macrophages, so for this study a TK deletion mutant virus (ΔTK) was developed using a Malawi Lil-20/1 strain partially adapted to grow in Vero cells, although still fully virulent in domestic pigs [23]. Animals inoculated IM with 10^4 TCID$_{50}$ ΔTK had 65% survival rate showing a transient period of fever without any other clinical signs of ASF. The surviving animals were IM challenged (performed at day 53 pv) with 10^4 TCID of the Malawi isolate remaining clinically normal.

The *9GL* gene [24], with similarity to yeast *ERV1* which is involved in oxidative phosphorylation and in cell growth, is highly conserved across isolates suggesting the possibility of a functional conserved genetic role in ASFV. This gene was first deleted in the highly virulent Malawi Lil-20/1strain, the resulting Δ9GL strain had a decreased ability to replicate in swine macrophages and, when examined under electron microscopy, it was observed that some of the virus particles had an altered morphology [24]. Importantly, Δ9GL was completely attenuated when IM inoculated in pigs in a range of 10^2-10^6 TCID$_{50}$ showing the animals just a transient mild rise in body temperature and a reduced viremia lasting 3-4 weeks. Importantly at 42 days pv, a challenge with 10^4 TCID$_{50}$ of parental highly virulent Malawi Lil-20/1 administrated IM resulted in complete protection with the exception of a transient rise in body temperature in animals inoculated with low doses of Δ9GL.

Therefore, these four genes (*NL-S*, *UK*, *TK*, and *9GL*) were identified as the first determinants of virulence since, when deleted in the tested ASFV isolate, produced a drastic reduction in the virulence of the parental virus in domestic swine. Importantly, all four attenuated strains developed (ΔNL-S, ΔUK, ΔTK, and Δ9GL), when used as experimental vaccines, efficiently protected pigs from disease produced by the challenge with the corresponding parental virus. Interestingly, although these genes were quite conserved across isolates, deletion of the same gene from the genome of different ASFV strains not always produced the same results in terms of virulence attenuation and protection efficacy as described in Section 4.

Published by Burleigh Dodds Science Publishing Limited, 2024.

Over the next several years, there were no new determinants of virulence discovered for ASFV, partially due to the decreased worldwide efforts due to ASFV being mostly restricted to Africa, and partially due to the difficulty in purification of recombinant ASFV. However, worldwide interest started to grow with the introduction of ASFV in the Republic of Georgia in 2007 and continues to grow as this outbreak has now reached pandemic proportions due to the continued spread in Europe and across most of Asia. In Section 4, we will describe the efforts to make a vaccine for this pandemic strain. However, there was continued research on other ASFV strains that are worth noting in this section.

The discovery of DP148R as an ASFV determinant of virulence was determined using the virulent ASFV isolate Benin 97/1, producing the BeninΔDP148R virus [25]. BeninΔDP148R had similar growth in primary swine macrophages as its parental Benin 97/1 strain. However, it significantly reduced virulence in domestic pigs. Animals IM inoculated with 10^3 HAD_{50} of BeninΔDP148R virus only developed a transient rise of body temperature without additional signs related to ASF. The virus was detected in blood for approximately 3 weeks in animals that received BeninΔDP148R. In this study, a booster was also given 21 days after the first dose, and animals were challenged IM with 10^4 HAD_{50} of parental virulent virus 3 weeks after the booster. Challenged animals survived showing some of them only a transitory period of fever.

Another determinant of virulence that was discovered in the Ba71 ASFV strain was CD2v (EP402R) [26]. Recombinant *BA71ΔCD2v* was developed by replacing the *EP402R* gene with a Lac repressor cassette under the control of the ASFV early/late promoter pU104 and the βGus reporter gene. This recombinant virus was purified using COS-1 cells, which were also used in the production of virus stocks. Pigs IM infected with 10^4 or 10^6 PFU of BA71ΔCD2v remain clinically normal with just a transient rise in body temperature and, importantly, were protected when IM challenged 24 days later with 10^3 HAD_{50} of virulent BA71 virus. Interestingly, when the BA71ΔCD2v was used to test against an IM challenge with 20 LD of the heterologous Georgia 2007/1 isolate all animals survived the challenge. In another virus, isolates deletion of *CD2v* gene has not been associated with similar reductions in virulence. Therefore, deletion of *CD2v* gene in historic isolates, such as Malawi Lil-20/1 [27], or in recent isolates, such as Georgia 2007 [28] and CN/GS/2018 [29], did not cause any decrease in virus virulence of the parental virus. It is likely that the specific genetic characteristics of the BA71 strain that was used as a parental virus resulted in the deletion of the *CD2v* gene to provoke a complete loss of virulence in BA71ΔCD2v [26].

Published by Burleigh Dodds Science Publishing Limited, 2024.

4   Case study: development of an African swine fever virus vaccine for the new emerging strain African swine fever virus Georgia

The introduction of ASFV into the Republic of Georgia in 2007 which quickly spread into surrounding countries brought attention back to the need for an ASF vaccine. In our research program, we first targeted the classical determinants of virulence that were identified in earlier isolates of ASF such as *NL-S*, *UK*, *TK*, and *9GL*. The first target was to delete NL-S from the Georgia2010 isolate. Deletion of this gene was the first clue that deletion of determinants of virulence in one isolate of ASFV may not have the same effect in other isolates, as deletion of the *NL-S* gene from the highly virulent ASFV Georgia2010 isolate did not attenuate the virus [30] as it did in the *E70* isolate where complete attenuation occurred [21]. Animals IM inoculated with 10^4 HAD_{50} either succumbed to ASF with an acute form of the disease or survived with delayed and subclinical disease. The next attempt to delete the gene *UK* resulted in a similar situation where when deleted in Georgia2010 isolate its virulence was not reduced. An IM inoculation of 10^4 HAD_{50} of ASFV-G-ΔUK in swine produced an indistinguishable disease from that caused by the parental Georgia2010 virus [30], unlike the attenuation observed in the E70 isolate when the gene *UK* was deleted [22].

The deletion of the *TK* gene in a virulent strain of ASFV Georgia adapted to replicate in Vero cells had a different outcome since the virus produced, ASFV-G/V-ΔTK, when IM inoculated in pigs up to 10^6 $TCID_{50}$ did not produce any signs of ASF disease. However, none of the animals that received ASFV-G/V-ΔTK obtained any immunity to ASF as animals were not protected when challenged with the virulent parental Georgia strain [31].

The deletion of the *9GL* gene in the ASFV isolates mentioned prior was also performed in the Pretoriuskop/96/4 (Pret4) with similar results in terms of attenuation [32] than in the Malawi Lil-20/1 isolate [24], making the *9GL* gene a promising target for deletion in ASFV. However, the deletion of 9GL in the ASFV Georgia 2007 isolate did not have the same degree of attenuation as observed in other isolates [33]. At low doses, up to 10^3 HAD IM inoculated, ASFV-G-Δ9GL remained fully attenuated and the inoculated animals were protected against challenges with the virulent parental strain of ASFV. However, when the dosage was increased, pigs receiving 10^4 HAD developed a fatal acute disease. These results indicate that ASFV-G-Δ9GL retains residual virulence that is not seen in Malawi Lil-20/1, which remains completely attenuated even when IM is administered at a dose as high as 10^6 HAD [24]. Despite the appearance of residual virulence, ASFV-G-Δ9GL, when used at sub-lethal doses (10^2 or 10^3 HAD_{50}), did induce complete protection against the IM challenge with 10^3 HAD_{50} of the Georgia 2007 to isolate as early as 21 days pv. Challenged animals were clinically normal presenting, most of them, absence of replication of the challenge virus. Therefore, ASFV-G-Δ9GL was one of the first recombinant attenuated viruses reported to induce protection against the current highly virulent pandemic strain of ASFV. The residual virulence observed at higher doses inhibited potential commercialization of this vaccine candidate.

Further attenuation of the ASFV-G-Δ9GL vaccine candidate was achieved by deletion of the UK gene. As mentioned earlier, the deletion of *UK* in the Georgia isolate does not decrease parental virus virulence [34], but when combined with the deletion of 9GL, resulting in ASFV-G-Δ9GL/ΔUK, this double deletion virus had a significant increase in attenuation compared with ASFV-G-Δ9GL [34]. While animals IM inoculated with 10^4 HAD_{50} of ASFV-G-Δ9GL developed a fatal form of ASF, those IM inoculated with up to 10^6 HAD_{50} of ASFV-G-Δ9GL/ΔUK remained clinically normal. ASFV-G-Δ9GL/ΔUK has a decreased ability to grow in swine macrophages when compared with the parental ASFV-G-Δ9GL but effectively protected pigs when inoculated with a dose of either 10^4 or 10^6 HAD_{50} against the IM challenge with 10^3 HAD_{50} of Georgia isolate.

As an alternative approach to deleting single genes to develop a live-attenuated vaccine, a group of genes was deleted that belonged to the multi-gene family of genes. Six of these genes were removed from the Georgia 2007 isolate, resulting in the recombinant virus ASFV-G-ΔMGF which lacks genes *MGF505-1R, MGF360-12L, MGF360-13L, MGF360-14L, MGF505-2R,* and *MGF505-3R* [35]. These genes were chosen as they have been associated with host range specificity and the down-modulation of the innate immune response [36, 37] and because deletions in this area of the genome have been found in naturally attenuated field isolates and in cell culture-adapted strains [13, 38, 39]. In primary swine macrophages, ASFV-G-ΔMGF replicates similarly to the parental virus. IM inoculation of up to 10^4 HAD_{50} ASFV-G-ΔMGF the swine remained healthy, without signs of the disease. Importantly, when animals receiving doses of 10^2 HAD_{50} or higher were challenged with virulent parental Georgia 2007 strain, no signs of the disease were observed, although a proportion of these animals harbored the challenge virus [35]. After this discovery, another group made a recombinant virus, HLJ/18-6GD, that harbors a similar gene deletion using a parental virus of the highly virulent Chinese field isolate HLJ/18 [29] confirming results obtained with ASFV-G-ΔMGF. Animals inoculated IM 10^4 HAD_{50} or lower doses remained clinically normal and when these animals were challenged with 200 Pig Lethal Dose$_{50}$ (PLD$_{50}$) the animals remained clinically normal. Reversion to virulence studies was performed using *HLJ/18-6GD* showing the presence of clinical signs associated with the disease observed in one of the animals in both passages 5 and 6 groups, which suggests that this virus may not be genetically stable [29]. However, no information was provided regarding the genomic changes associated with the disease observed [29]. The same research group presented an improved version of the *HLJ/18-6GD* virus by additionally deleting the CD2-like gene. The resulting virus *HLJ/18-7GD* presented the same protective efficacy as *HLJ/18-6GD* and remained phenotypically stable in reversion to virulence studies [29]; however, CD2 alone does not affect virus virulence in this isolate [29] or in the Georgia 2007 isolate [28]. We also determined that the addition of CD2 to a different vaccine strain ASFV-G-Δ9GL reduced the efficacy of the parental vaccine [40].

The discovery of the virulence determinant I177L [41] was first performed by deleting this gene in the genome of the ASFV-Georgia 2007 isolate. The deletion of this previously uncharacterized and highly conserved gene produced a reduction of the virus growth in cultures of swine macrophages. The recombinant virus lacking this gene, *ASFV-G-ΔI177L*, presented a drastic decrease in virulence when inoculated in pigs even when animals received up to 10^6 HAD$_{50}$ IM. All animals remained clinically normal without even presenting a transient rise in body temperature, an observation rarely seen with other recombinant viruses tested at this dosage. Importantly, animals receiving even a single low dose of 10^2 HAD$_{50}$ ASFV-G-ΔI177L doses were completely protected against the challenge IM with 10^2 HAD$_{50}$ of Georgia 2007 virus at 28 days pv. Remarkably, animals vaccinated with doses of 10^4 HAD$_{50}$ or higher of ASFV-G-ΔI177L developed sterile immunity against the challenge virus. The characteristics of ASFV-G-ΔI177L as a lead vaccine candidate were further tested to determine if the vaccine would offer protection to a recent Asian outbreak strain from Vietnam, TTKN/ASFV/DN/2019 [42]. TTKN/ASFV/DN/2019 is a derivative of the Georgia 2007 isolate that has evolved in the field for at least 12 years. This study determined that ASFV-G-ΔI177L is able to induce protection both against TTKN/ASFV/DN/2019 and the Georgia 2007 strain [42] with similar efficacy. To further test the range of this vaccine experiments were also conducted to compare the effectiveness of this vaccine in pigs with both, European (Yorkshire/Landrace crossbreed) and Vietnamese genetic backgrounds (Mong Cai breed). No differences were observed between either pig's breed, with vaccinated animals being protected against the virulent challenge of TTKN/ASFV/DN/2019.

To explore the possibility of using ASFV-G-ΔI177L as an oral vaccine, the virus was administered by the oronasal route. Oronasally inoculated animals with 10^6 HAD$_{50}$ of ASFV-G-ΔI177L were IM challenged 28 days later with 10^2 HAD$_{50}$ of the virulent Georgia isolate were fully protected and had similar results to IM-inoculated animals not showing clinical signs associated with ASF [43].

Despite having multiple live-attenuated vaccine platforms, we continue to identify new determinants of virulence, as new emerging strains may not achieve the same degree of attenuation by introducing the same gene deletions, as we observed with the deletion of *9GL* gene in ASFV Georgia isolate when compared with results obtained with historical isolates. The deletion of the highly conserved gene *A137R* resulted in recombinant virus ASFV-G-ΔA137R [44], which had a moderate replication decrease in primary swine macrophage cultures. When tested using an IM dose of 10^2 HAD$_{50}$, ASFV-G-ΔA137R was significantly attenuated in swine with animals remaining clinically healthy during the 28-day observational period. When IM was challenged with 10^2 HAD$_{50}$ of ASFV-G all ASFV-G-ΔA137R inoculated animals were protected with no evidence of replication of challenge virus.

In an effort to discover a candidate gene that could be used as a serological marker, ASFV gene *E184L* was deleted to produce a recombinant virus ASFV-G-ΔE184L [45]. In animals inoculated IM with 10^2 HAD$_{50}$ of ASFV-G-ΔE184L 60% of animals survived the infection just presenting a transient period of fever while the other 40% experienced a significantly delayed presentation of a fatal clinical disease. Interestingly, all animals surviving ASFV-G-ΔE184L infection were protected when IM was challenged with 10^2 HAD$_{50}$ of parental Georgia isolate. Importantly, animals surviving ASFV-G-ΔE184L infection did not develop antibodies to the *E184L* gene. Further studies demonstrated that the protein encoded by *E184L* gene is highly immunogenic inducing a readily detected antibody response during the infection with other ASFV strains. Therefore, deletion of the *E184L* gene could be used as a serological marker to differentiate between infected and vaccinated pigs, if *E184L* could be added to a vaccine candidate. However, when *E184L* was added to the ASFV-G-ΔMGF vaccine candidate, the efficacy of the parental vaccine was lost [45].

One of the problems with all the ASFV experimental vaccines produced using Georgia isolate or its derivatives as a backbone are that the experimental vaccines require primary swine macrophages for growth. Swine macrophages have to be freshly isolated from donor swine using a time-consuming process, and face additional regulatory requirements to produce a vaccine. To overcome this hurdle a large-scale effort was made to test different cell lines for growth of the ASFV-G-ΔI177L vaccine. Adapting ASFV-G-ΔI177L to grow in an established swine cell line, PIPEC cells (Plum Island Porcine Epithelial Cells), [46] resulted in an additional deletion in the left variable region (ΔLVR). This adapted virus, ASFV-G-ΔI177L/ΔLVR, in challenge studies performed in domestic pigs, maintained the same level of attenuation, immunogenic characteristics, and protective efficacy as ASFV-G-ΔI177L [46]. The discovery of ASFV-G-ΔI177L/ΔLVR is very promising as production on an industry-level scale is easier in an established cell line and overcomes some of the regulatory hurdles associated with using primary swine macrophages, as required for other vaccine candidates.

5 Other advances in African swine fever virus discovery of determinants of virulence

The previous case study section highlights the advances that were made in our group to develop an ASFV vaccine for the pandemic strain, however, other groups have also identified different determinants of virulence for the pandemic strain of ASFV Georgia 2007. The summary of these individual advances is as follows.

ASFV gene *MGF-110-9L* belongs to the MGF 110 Family in ASFV and, although the function of this gene is largely unknown, it was recently shown to be involved in disease production [47]. A recombinant virus ASFV-ΔMGF110-9L, with a deletion of the MGF-110-9L gene, developed from the highly virulent ASFV Chinese isolate CN/GS/2018 (a derivative of the Georgia isolate) had decreased replication in primary swine macrophage cell cultures. ASFV-ΔMGF110-9L when tested in swine presented a partial reduction in virulence compared with its parental virus with 60% of animals IM inoculated with 10 HAD$_{50}$ of ASFV-ΔMGF110-9L surviving the infection with only a transitory period of fever and low titers in blood samples. No data was presented regarding the ability of the ASFV-ΔMGF110-9L in inducing protection against the challenge of the virulent parental virus [47].

The individual deletion of another member of the MGF, ASFV gene *MGF-505-7R*, was tested for virulence in pigs [48]. The role of the *MGF-505-7R* gene was determined in this study to be in the down-modulation of the innate immune response. The recombinant virus (ASFV-ΔMGF505-7R) was based in the virulent field isolate CN/GS/2018 harboring a single deletion of the *MGF-505-7R* gene. Pigs IM inoculated with 10 HA$_{50}$ of the recombinant virus presented an absence of clinical signs associated with the disease during the 3 weeks observation period. ASFV-ΔMGF505-7R replication in the blood and tissues of inoculated pigs was decreased in comparison with parental ASF; however, higher doses or challenge studies were not performed [48].

Another determinant of virulence that was deleted from a virulent ASFV strain CN/GS/2018 is ASFV gene *MGF360-9L*. This gene has also been shown to be involved in the downregulation of interferon expression [49]. Recombinant virus ASFV-Δ360-9L, with this gene deleted, had a minimal reduction in growth in primary swine macrophages and had partial attenuation in pigs where animals IM infected with 1 HAD$_{50}$ 25% presented a fatal disease, transient clinical disease in another 20% and 55% remained clinically normal. Higher doses were not tested in this study [49]; however, due to the clinical observation at this low dose it is likely that higher doses would have more pounced clinical effects.

These recent studies involving ASFV-Δ9L [47], ASFV-Δ7R [50], and ASFV-Δ360-9L [49] did not provide any data regarding the capacity of the recombinant viruses to protect inoculated animals against the parental virus challenge and in all of these studies only a very low dose of virus (1-10 HAD) was used to test attenuation when inoculated in animals. Higher dose studies will be needed to be performed along with challenge studies to fully evaluate these new genetic determinants of virulence.

The deletion of I226R, a previously uncharacterized ASFV gene, demonstrated that this gene is a determinant of virulence in the SY18 isolate [51]. Deletion of the gene in the genome of the virulent SY18 (SY18 ΔI226R) was tested in pigs inoculated IM with doses as high as 10^7 $TCID_{50}$. All of these animals remained clinically normal showing reduced levels of viremia. After IM challenge using the parental SY18 strain at a dose of 10^4 $TCID_{50}$ at 21 days pv all the challenged pigs survived, without ASF clinical signs.

A recombinant virus with the deletion of five contiguously located genes (L7L-L11L) in the genome of the virulent strain SY18, SY18ΔL7-11, is able to replicate at similar levels as the parental SY18 isolate in primary swine macrophages. When IM inoculated in swine at doses of 10^3 or 10^6 HAD_{50} showed 90% of the swine survived with most of the animals showing a transitory fever and, in some cases, mild signs of disease. All surviving animals were protected from the IM challenge with 10^3 HAD_{50} of parental SY18 virus with only a few of them having a transitory rise in body temperature [52].

Recently, it was shown that the ASFV gene *I267L* inhibits the RNA Pol-III-RIG-I-mediated innate antiviral responses [53] when deleted from the virulent isolate CN/GS/2018 (ASFVΔI267L). When pigs were inoculated IM with 10 HAD_{50} of ASFVΔI267L, 80% of them exhibited mild clinical signs but they survived the 3-week observation period, while the rest presented a protracted and fatal disease. Interestingly, another group deleted *I267L* from a different virulent Chinese field isolate SY18, producing the recombinant SY18ΔI267L strain. In this case, when SY18ΔI267L was IM inoculated at doses of $10^{2.0}$ $TCID_{50}$ a severe and lethal disease in all animals occurred [54]. It is unclear why the deletion in two similar viruses had different results but each study had differences in the inoculated dose described or there could be other genetic differences between the two different isolates CN/GS/2018 and SY18 even though both are derivatives of the Georgia outbreak virus.

Another recent report that deleted ASFV gene *I1226R* [45] showed that deletion of *I226R* from Chinese isolate SY18 a virulent strain of ASFV (SY18ΔI226R) Pigs inoculated IM with doses as high as 10^7 $TCID_{50}$ remained clinically normal and had reduced viremia. After the IM challenge with 10^4 $TCID_{50}$ of the parental strain at 21 days pv, all the challenged pigs survived with no development of ASF disease showing that I226R caused a severe decrease in virulence and could be a promising vaccine candidate.

Combining deletions in *CD2* and *UK* genes resulted in the recombinant virus ASFV-SY18-ΔCD2v/UK [55]. When ASFV-SY18-ΔCD2v/UK was tested in animals receiving IM 10^4 $TCID_{50}$ the inoculated animals did not present clinical signs other than a transient fever that is associated with ASF. The challenge, at 28 days pv, with 10^4 TCID$_{50}$ of the parental virus resulted in all of the ASFV-SY18-ΔCD2v/UK-inoculated pigs surviving the challenge. Interestingly, a similar recombinant virus, HLJ/18-9GL&UK, harboring the same deletions of both the *CD2* and *UK* genes, had residual virulence producing a lethal disease in 50% of animals IM inoculated with 10^4 TCID$_{50}$ [29]. Again, as in the deletion of I267L described above, differences in reproducing results obtained by deleting the same genes in a very similar genetic background could perhaps be due to small changes in these two genetic backgrounds of the parental viruses.

6   Conclusion

The first experimental live-attenuated vaccines have been made for ASF, the studies performed, and the characteristics of the different vaccines vary. As there is no standardized challenge model for ASFV vaccines, differences between laboratories make it hard to directly compare results. For example, the amount of IM challenge or dose range of the vaccine candidates varies between studies. However, there are several promising vaccines that have been made using homologous backbones to virulent strains, including in recent literature vaccines to the current pandemic strain of ASFV that is circulating in Europe and Asia, and more recently in the Dominican Republic and Haiti. As this pandemic strain has been circulating with minimal genomic changes for 15 years and in this time has continued to spread to new countries each year, a commercialized vaccine is needed to control this pandemic. At the time of this writing, the first generation of a live-attenuated vaccine ASFV-G-ΔI177L is the most advanced and has obtained approval for commercial production and use in Vietnam. This is an important advance in ASF vaccines showing that attenuated ASF viruses have the potential to become safe vaccines. However, vaccine strains ASFV-G-Δ9GL/UK, ASFV-G-ΔMGF, HLJ/18-7GD are being further developed. A likely second-generation live-attenuated vaccine candidate for ASFV that is in the process of being further developed would ASFV-G-ΔI177LΔLVR that does not rely on primary swine macrophages.

7   Future trends in research

Prior to the initial outbreak of AFSV Georgia in 2007, this virus was not well studied, and no attempts were made for a vaccine of this genotype. Once the outbreak started to spread it became clear that an effective vaccine was needed. However, it was quickly learned that historical information for genetic determinants of virulence obtained from other virus isolates did not reproduce the same results obtained when deleted in the ASFV Georgia strain, as was described, for example, using the 9GL deletion. Due to this fact, it is important that the discovery of additional determinants of virulence in ASFV is continued, as this knowledge may be needed in future outbreaks.

Published by Burleigh Dodds Science Publishing Limited, 2024.

Despite having some very good live-attenuated ASFV experimental vaccines for the current pandemic strain, with ASFV-G-ΔI177L being commercially produced in Vietnam, currently only has approval for use in Vietnam. Even with this new milestone achieved for a live-attenuated ASF vaccine, there is still a need to develop either vector-based or subunit vaccines for ASFV so that vaccines could be more widely used in high-risk but uninfected countries, to help control the further spread of the outbreaks. Another reason is that live-attenuated vaccines are very time-consuming to produce and face higher regulatory hurdles should a new outbreak strain be discovered that the current live-attenuated vaccines do not offer protection against. Some subunits or vectored vaccines could offer a very quick turnaround time in a future outbreak situation. However, one of the current gaps in ASFV research is the information on what ASFV proteins could be used as successful candidates for a vectored or subunit vaccine.

8   Where to look for further information

- The Global African swine fever Research Alliance (GARA) is the link to all ASFV research labs worldwide and maintains a webpage with current GAP analysis reports made by these research labs. Information is available here: https://www.ars.usda.gov/gara/.
- The World Organization of Animal Health (OiE) maintains a webpage with information on ASFV available here: https://www.oie.int/en/disease/african-swine-fever/.
- Center of Excellence for African Swine Fever Genomics maintains a webpage with information on all individual ASFV proteins and is available here: https://www.ASFVgenomics.com.

9   References

1. Sunwoo, S. Y., Pérez-Núñez, D., Morozov, I., Sánchez, E. G., Gaudreault, N. N., Trujillo, J. D., Mur, L., Nogal, M., Madden, D., Urbaniak, K., Kim, I. J., Ma, W., Revilla, Y. and Richt, J. A. DNA-protein vaccination strategy does not protect from challenge with African swine fever virus Armenia 2007 strain. *Vaccines (Basel)* 2019 7(1): 12.
2. Argilaguet, J. M., Pérez-Martín, E., López, S., Goethe, M., Escribano, J. M., Giesow, K., Keil, G. M. and Rodríguez, F. BacMam immunization partially protects pigs against sublethal challenge with African swine fever virus. *Antiviral Res.* 2013 98(1): 61–5.
3. Lokhandwala, S., Petrovan, V., Popescu, L., Sangewar, N., Elijah, C., Stoian, A., Olcha, M., Ennen, L., Bray, J., Bishop, R. P. and Waghela, S. D. Adenovirus-vectored African swine fever virus antigen cocktails are immunogenic but not protective against intranasal challenge with Georgia. *Vet. Microbiol.* 2019/2007 235: 10–20.
4. Lokhandwala, S., Waghela, S. D., Bray, J., Martin, C. L., Sangewar, N., Charendoff, C., Shetti, R., Ashley, C., Chen, C. H., Berghman, L. R., Mwangi, D., Dominowski, P. J., Foss, D. L., Rai, S., Vora, S., Gabbert, L., Burrage, T. G., Brake, D., Neilan, J. and Mwangi, W. Induction of robust immune responses in Swine by using a cocktail of

Published by Burleigh Dodds Science Publishing Limited, 2024.

adenovirus-vectored African swine fever virus antigens. *Clin. Vaccine Immunol.* 2016 23(11): 888-900.

5. Cadenas-Fernandez, E., Sánchez-Vizcaíno, J. M., Kosowska, A., Rivera, B., Mayoral-Alegre, F., Rodríguez-Bertos, A., Yao, J., Bray, J., Lokhandwala, S., Mwangi, W. and Barasona, J. A. Adenovirus-vectored African swine fever virus antigens cocktail is not protective against virulent Arm07 isolate in Eurasian wild boar. *Pathogens* 2020 9(3): 171.

6. Netherton, C. L., Goatley, L. C., Reis, A. L., Portugal, R., Nash, R. H., Morgan, S. B., Gault, L., Nieto, R., Norlin, V., Gallardo, C., Ho, C. S., Sánchez-Cordón, P. J., Taylor, G. and Dixon, L. K. Identification and immunogenicity of African swine fever virus antigens. *Front. Immunol.* 2019 10: 1318.

7. Goatley, L. C., Reis, A. L., Portugal, R., Goldswain, H., Shimmon, G. L., Hargreaves, Z., Ho, C. S., Montoya, M., Sánchez-Cordón, P. J., Taylor, G., Dixon, L. K. and Netherton, C. L. A pool of eight virally vectored African swine fever antigens protect pigs against fatal disease. *Vaccines (Basel)* 2020 8(2): 234.

8. Lokhandwala, S., Waghela, S. D., Bray, J., Sangewar, N., Charendoff, C., Martin, C. L., Hassan, W. S., Koynarski, T., Gabbert, L., Burrage, T. G., Brake, D., Neilan, J. and Mwangi, W. Adenovirus-vectored novel African swine fever virus antigens elicit robust immune responses in swine. *PLoS ONE* 2017 12(5): e0177007.

9. Detray, D. E. Persistence of viremia and immunity in African swine fever. *Am. J. Vet. Res.* 1957 18(69): 811-6.

10. Malmquist, W. A. Serologic and immunologic studies with African swine fever virus. *Am. J. Vet. Res.* 1963 24: 450-9.

11. Mebus, C. A. and Dardiri, A. H. Western hemisphere isolates of African swine fever virus: Asymptomatic carriers and resistance to challenge inoculation. *Am. J. Vet. Res.* 1980 41(11): 1867-9.

12. Leitao, A., Cartaxeiro, C., Coelho, R., Cruz, B., Parkhouse, R. M. E., Portugal, F. C., Vigário, J. D. and Martins, C. L. V. The non-haemadsorbing African swine fever virus isolate ASFV/NH/P68 provides a model for defining the protective anti-virus immune response. *J. Gen. Virol.* 2001 82(3): 513-23.

13. Krug, P. W., Holinka, L. G., O'Donnell, V., Reese, B., Sanford, B., Fernandez-Sainz, I., Gladue, D. P., Arzt, J., Rodriguez, L., Risatti, G. R. and Borca, M. V. The progressive adaptation of a Georgian isolate of African swine fever virus to Vero cells leads to a gradual attenuation of virulence in swine corresponding to major modifications of the viral genome. *J. Virol.* 2015 89(4): 2324-32.

14. Sereda, A. D., Balyshev, V. M., Kazakova, A. S., Imatdinov, A. R. and Kolbasov, D. V. Protective properties of attenuated strains of African swine fever virus belonging to Seroimmunotypes I-VIII. *Pathogens* 2020 9(4): 274.

15. Boinas, F. S., Hutchings, G. H., Dixon, L. K. and Wilkinson, P. J. Characterization of pathogenic and non-pathogenic African swine fever virus isolates from Ornithodoros erraticus inhabiting pig premises in Portugal. *J. Gen. Virol.* 2004 85(8): 2177-87.

16. Oura, C. A. L., Denyer, M. S., Takamatsu, H. and Parkhouse, R. M. E. *In vivo* depletion of CD8+ T lymphocytes abrogates protective immunity to African swine fever virus. *J. Gen. Virol.* 2005 86(9): 2445-50.

17. King, K., Chapman, D., Argilaguet, J. M., Fishbourne, E., Hutet, E., Cariolet, R., Hutchings, G., Oura, C. A., Netherton, C. L., Moffat, K., Taylor, G., Le Potier, M. F., Dixon, L. K. and Takamatsu, H. H. Protection of European domestic pigs from virulent African isolates of African swine fever virus by experimental immunisation. *Vaccine* 2011 29(28): 4593-600.

18. Gallardo, C., Soler, A., Rodze, I., Nieto, R., Cano-Gómez, C., Fernandez-Pinero, J. and Arias, M. Attenuated and non-haemadsorbing (non-HAD) genotype II African swine fever virus (ASFV) isolated in Europe, Latvia 2017. *Transbound. Emerg. Dis.* 2019 66(3): 1399–404.

19. Gallardo, C., Sánchez, E. G., Pérez-Núñez, D., Nogal, M., de León, P., Carrascosa, Á. L., Nieto, R., Soler, A., Arias, M. L. and Revilla, Y. African swine fever virus (ASFV) protection mediated by NH/P68 and NH/P68 recombinant live-attenuated viruses. *Vaccine* 2018 36(19): 2694–704.

20. Abrams, C. C., Goatley, L., Fishbourne, E., Chapman, D., Cooke, L., Oura, C. A., Netherton, C. L., Takamatsu, H. H. and Dixon, L. K. Deletion of virulence associated genes from attenuated African swine fever virus isolate OUR T88/3 decreases its ability to protect against challenge with virulent virus. *Virology* 2013 443(1): 99–105.

21. Zsak, L., Lu, Z., Kutish, G. F., Neilan, J. G. and Rock, D. L. An African swine fever virus virulence-associated gene NL-S with similarity to the herpes simplex virus ICP34.5 gene. *J. Virol.* 1996 70(12): 8865–71.

22. Zsak, L., Caler, E., Lu, Z., Kutish, G. F., Neilan, J. G. and Rock, D. L. A nonessential African swine fever virus gene UK is a significant virulence determinant in domestic swine. *J. Virol.* 1998 72(2): 1028–35.

23. Moore, D. M., Zsak, L., Neilan, J. G., Lu, Z. and Rock, D. L. The African swine fever virus thymidine kinase gene is required for efficient replication in swine macrophages and for virulence in swine. *J. Virol.* 1998 72(12): 10310–5.

24. Lewis, T., Zsak, L., Burrage, T. G., Lu, Z., Kutish, G. F., Neilan, J. G. and Rock, D. L. An African swine fever virus ERV1-ALR homologue, 9GL, affects virion maturation and viral growth in macrophages and viral virulence in swine. *J. Virol.* 2000 74(3): 1275–85.

25. Reis, A. L., Goatley, L. C., Jabbar, T., Sanchez-Cordon, P. J., Netherton, C. L., Chapman, D. A. G. and Dixon, L. K. Deletion of the African swine fever virus gene DP148R does not reduce virus replication in culture but reduces virus virulence in pigs and induces high levels of protection against challenge. *J. Virol.* 2017 91(24).

26. Monteagudo, P. L., Lacasta, A., López, E., Bosch, L., Collado, J., Pina-Pedrero, S., Correa-Fiz, F., Accensi, F., Navas, M. J., Vidal, E., Bustos, M. J., Rodríguez, J. M., Gallei, A., Nikolin, V., Salas, M. L. and Rodríguez, F. BA71DeltaCD2: A new recombinant live attenuated African swine fever virus with cross-protective capabilities. *J. Virol.* 2017 91(21): e01058-17.

27. Borca, M. V., Carrillo, C., Zsak, L., Laegreid, W. W., Kutish, G. F., Neilan, J. G., Burrage, T. G. and Rock, D. L. Deletion of a CD2-like gene, 8-DR, from African swine fever virus affects viral infection in domestic swine. *J. Virol.* 1998 72(4): 2881–9.

28. Borca, M. V., O'Donnell, V., Holinka, L. G., Risatti, G. R., Ramirez-Medina, E., Vuono, E. A., Shi, J., Pruitt, S., Rai, A., Silva, E., Velazquez-Salinas, L. and Gladue, D. P. Deletion of CD2-like gene from the genome of African swine fever virus strain Georgia does not attenuate virulence in swine. *Sci. Rep.* 2020 10(1): 494.

29. Chen, W., Zhao, D., He, X., Liu, R., Wang, Z., Zhang, X., Li, F., Shan, D., Chen, H., Zhang, J., Wang, L., Wen, Z., Wang, X., Guan, Y., Liu, J. and Bu, Z. A seven-gene-deleted African swine fever virus is safe and effective as a live attenuated vaccine in pigs. *Sci. China Life Sci.* 2020 63(5): 623–34.

30. Ramirez-Medina, E., Vuono, E., O'Donnell, V., Holinka, L. G., Silva, E., Rai, A., Pruitt, S., Carrillo, C., Gladue, D. P. and Borca, M. V. Differential effect of the deletion of African swine fever virus virulence-associated genes in the induction of attenuation of the highly virulent Georgia strain. *Viruses* 2019 11(7): 599.

Published by Burleigh Dodds Science Publishing Limited, 2024.

31. Sanford, B., Holinka, L. G., O'Donnell, V., Krug, P. W., Carlson, J., Alfano, M., Carrillo, C., Wu, P., Lowe, A., Risatti, G. R., Gladue, D. P. and Borca, M. V. Deletion of the thymidine kinase gene induces complete attenuation of the Georgia isolate of African swine fever virus. *Virus Res.* 2016 213: 165-71.

32. Carlson, J., O'Donnell, V., Alfano, M., Velazquez Salinas, L., Holinka, L. G., Krug, P. W., Gladue, D. P., Higgs, S. and Borca, M. V. Association of the host immune response with protection using a live attenuated African swine fever virus model. *Viruses* 2016 8(10): 291.

33. O'Donnell, V., Holinka, L. G., Krug, P. W., Gladue, D. P., Carlson, J., Sanford, B., Alfano, M., Kramer, E., Lu, Z., Arzt, J., Reese, B., Carrillo, C., Risatti, G. R. and Borca, M. V. African swine fever virus Georgia 2007 with a deletion of virulence-associated gene 9GL (B119L), when administered at low doses, leads to virus attenuation in swine and induces an effective protection against homologous challenge. *J. Virol.* 2015 89(16): 8556-66.

34. O'Donnell, V., Risatti, G. R., Holinka, L. G., Krug, P. W., Carlson, J., Velazquez-Salinas, L., Azzinaro, P. A., Gladue, D. P. and Borca, M. V. Simultaneous deletion of the 9GL and UK genes from the African swine fever virus Georgia 2007 isolate offers increased safety and protection against homologous challenge. *J. Virol.* 2017 91(1): e01760-16.

35. O'Donnell, V., Holinka, L. G., Gladue, D. P., Sanford, B., Krug, P. W., Lu, X., Arzt, J., Reese, B., Carrillo, C., Risatti, G. R. and Borca, M. V. African swine fever virus Georgia isolate harboring deletions of MGF360 and MGF505 genes is attenuated in Swine and confers protection against challenge with virulent parental virus. *J. Virol.* 2015 89(11): 6048-56.

36. Burrage, T. G., Lu, Z., Neilan, J. G., Rock, D. L. and Zsak, L. African swine fever virus multigene family 360 genes affect virus replication and generalization of infection in Ornithodoros porcinus ticks. *J. Virol.* 2004 78(5): 2445-53.

37. Zsak, L., Lu, Z., Burrage, T. G., Neilan, J. G., Kutish, G. F., Moore, D. M. and Rock, D. L. African swine fever virus multigene family 360 and 530 genes are novel macrophage host range determinants. *J. Virol.* 2001 75(7): 3066-76.

38. Pires, S., Ribeiro, G. and Costa, J. V. Sequence and organization of the left multigene family 110 region of the Vero-adapted L60V strain of African swine fever virus. *Virus Genes* 1997 15(3): 271-4.

39. Boinas, F. S., Hutchings, G. H., Dixon, L. K. and Wilkinson, P. J. Characterization of pathogenic and non-pathogenic African swine fever virus isolates from Ornithodoros erraticus inhabiting pig premises in Portugal. *J. Gen. Virol.* 2004 85(8): 2177-87.

40. Gladue, D. P., O'Donnell, V., Ramirez-Medina, E., Rai, A., Pruitt, S., Vuono, E. A., Silva, E., Velazquez-Salinas, L. and Borca, M. V. Deletion of Cd2-like (CD2v) and C-type lectin-like (EP153R) genes from African swine fever virus Georgia-9GL abrogates its effectiveness as an experimental vaccine. Viruses. *Deletion CD* 2020 12(10).

41. Borca, M. V., Ramirez-Medina, E., Silva, E., Vuono, E., Rai, A., Pruitt, S., Holinka, L. G., Velazquez-Salinas, L., Zhu, J. and Gladue, D. P. Development of a highly effective African swine fever virus vaccine by deletion of the I177L gene results in sterile immunity against the current epidemic Eurasia strain. *J. Virol.* 2020 94(7): e02017-19.

42. Tran, X. H., Le, T. T. P., Nguyen, Q. H., Do, T. T., Nguyen, V. D., Gay, C. G., Borca, M. V. and Gladue, D. P. African swine fever virus vaccine candidate ASFV-G-Delta177L efficiently protects European and native pig breeds against circulating Vietnamese field strain. *Transbound. Emerg. Dis.* 2022 69: e497-e504.

Published by Burleigh Dodds Science Publishing Limited, 2024.

43. Borca, M. V., Ramirez-Medina, E., Silva, E., Vuono, E., Rai, A., Pruitt, S., Espinoza, N., Velazquez-Salinas, L., Gay, C. G. and Gladue, D. P. ASFV-G-I177L as an effective oral nasal vaccine against the Eurasia strain of Africa swine fever. *Viruses* 2021 13(5): 765.

44. Gladue, D. P., Ramirez-Medina, E., Vuono, E., Silva, E., Rai, A., Pruitt, S., Espinoza, N., Velazquez-Salinas, L. and Borca, M. V. Deletion of the A137R gene from the pandemic strain of African swine fever virus attenuates the strain and offers protection against the virulent pandemic virus. *J. Virol.* 2021 95(21): e0113921.

45. Ramirez-Medina, E., Vuono, E., Rai, A., Pruitt, S., Espinoza, N., Velazquez-Salinas, L., Pina-Pedrero, S., Zhu, J., Rodriguez, F., Borca, M. V. and Gladue, D. P. Deletion of E184L, a putative DIVA target from the pandemic strain of African swine fever virus, produces a reduction in virulence and protection against virulent challenge. *J. Virol.* 2022 96(1): e0141921.

46. Borca, M. V., Rai, A., Ramirez-Medina, E., Silva, E., Velazquez-Salinas, L., Vuono, E., Pruitt, S., Espinoza, N. and Gladue, D. P. A cell culture-adapted vaccine virus against the current African swine fever virus pandemic strain. *J. Virol.* 2021 95(14): e0012321.

47. Li, D., Liu, Y., Qi, X., Wen, Y., Li, P., Ma, Z., Liu, Y., Zheng, H. and Liu, Z. African swine fever virus MGF-110-9L-deficient mutant has attenuated virulence in pigs. *Virol. Sin.* 2021 36(2): 187–95.

48. Li, D., Yang, W., Li, L., Li, P., Ma, Z., Zhang, J., Qi, X., Ren, J., Ru, Y., Niu, Q., Liu, Z., Liu, X. and Zheng, H. African swine fever virus MGF-505-7R negatively regulates cGAS-STING-mediated signaling pathway. *J. Immunol.* 2021 206(8): 1844–57.

49. Zhang, K., Yang, B., Shen, C., Zhang, T., Hao, Y., Zhang, D., Liu, H., Shi, X., Li, G., Yang, J., Li, D., Zhu, Z., Tian, H., Yang, F., Ru, Y., Cao, W. J., Guo, J., He, J., Zheng, H. and Liu, X. MGF360-9L is a major virulence factor associated with the African swine fever virus by antagonizing the JAK/STAT signaling pathway. *mBio* 2022 13(1): e0233021.

50. Li, D., Zhang, J., Yang, W., Li, P., Ru, Y., Kang, W., Li, L., Ran, Y. and Zheng, H. African swine fever virus protein MGF-505-7R promotes virulence and pathogenesis by inhibiting JAK1- and JAK2-mediated signaling. *J. Biol. Chem.* 2021 297(5): 101190.

51. Zhang, Y., Ke, J., Zhang, J., Yang, J., Yue, H., Zhou, X., Qi, Y., Zhu, R., Miao, F., Li, Q., Zhang, F., Wang, Y., Han, X., Mi, L., Yang, J., Zhang, S., Chen, T. and Hu, R. African swine fever virus bearing an I226R gene deletion elicits robust immunity in pigs to African swine fever. *J. Virol.* 2021 95(23): e0119921.

52. Zhang, J., Zhang, Y., Chen, T., Yang, J., Yue, H., Wang, L., Zhou, X., Qi, Y., Han, X., Ke, J., Wang, S., Yang, J., Miao, F., Zhang, S., Zhang, F., Wang, Y., Li, M. and Hu, R. Deletion of the L7L-L11L genes attenuates ASFV and induces protection against homologous challenge. *Viruses* 2021 13(2): 255.

53. Ran, Y., Li, D., Xiong, M. G., Liu, H. N., Feng, T., Shi, Z. W., Li, Y. H., Wu, H. N., Wang, S. Y., Zheng, H. X. and Wang, Y. Y. African swine fever virus I267L acts as an important virulence factor by inhibiting RNA polymerase III-RIG-I-mediated innate immunity. *PLoS Pathog.* 2022 18(1): e1010270.

54. Zhang, Y., Ke, J., Zhang, J., Yue, H., Chen, T., Li, Q., Zhou, X., Qi, Y., Zhu, R., Wang, S., Miao, F., Zhang, S., Li, N., Mi, L., Yang, J., Yang, J., Han, X., Wang, L., Li, Y. and Hu, R. I267L is neither the virulence- nor the replication-related gene of African swine fever virus and its deletant is an ideal fluorescent-tagged virulence strain. *Viruses* 2021 14(1): 53.

55. Teklue, T., Wang, T., Luo, Y., Hu, R., Sun, Y. and Qiu, H. J. Generation and evaluation of an African swine fever virus mutant with deletion of the CD2v and UK genes. *Vaccines (Basel)* 2020 8(4): 763.

Published by Burleigh Dodds Science Publishing Limited, 2024.

Index

9 781786 768612